FSA Subject Test

Mathematics Grade 7

Student Practice Workbook

+ Two Full-Length Florida FSA Math Tests

Math Notion

www.MathNotion.com

FSA Subject Test Mathematics Grade 7

FSA Subject Test Mathematics Grade 7

Published in the United State of America By

The Math Notion

Web: WWW.MathNotion.com

Email: info@Mathnotion.com

Copyright © 2021 by the Math Notion. All rights reserved. No part of this publication may be reproduced, stored in a retrieval system, or transmitted in any form or by any means, electronic, mechanical, photocopying, recording, scanning, or otherwise, except as permitted under Section 107 or 108 of the 1976 United States Copyright Ac, without permission of the author.

All inquiries should be addressed to the Math Notion.

ISBN: 978-1-63620-063-7

FSA Subject Test Mathematics Grade 7

The Math Notion

Michael Smith has been a math instructor for over a decade now. He launched the Math Notion. Since 2006, we have devoted our time to both teaching and developing exceptional math learning materials. As a test prep company, we have worked with thousands of students. We have used the feedback of our students to develop a unique study program that can be used by students to drastically improve their math scores fast and effectively. We have more than a thousand Math learning books including:

– SAT Math Prep

– ACT Math Prep

– SSAT/ISEE Math Prep

– Accuplacer Math Prep

– Common Core Math Prep

–many Math Education Workbooks, Study Guides, Practice and Exercise Books

As an experienced Math test preparation company, we have helped many students raise their standardized test scores—and attend the colleges of their dreams: We tutor online and in person, we teach students in large groups, and we provide training materials and textbooks through our website and through Amazon.

You can contact us via email at:

info@Mathnotion.com

FSA Subject Test Mathematics Grade 7

Get the Targeted Practice You Need to Ace the FSA Florida Math Test!

FSA Subject Test Mathematics Grade 7 includes easy-to-follow instructions, helpful examples, and plenty of math practice problems to assist students to master each concept, brush up their problem-solving skills, and create confidence.

The FSA math practice book provides numerous opportunities to evaluate basic skills along with abundant remediation and intervention activities. It is a skill that permits you to quickly master intricate information and produce better leads in less time.

Students can boost their test-taking skills by taking the book's two practice FSA Math exams. All test questions answered and explained in detail.

Important Features of the 7th grade FSA Math Book:

- A **complete review** of FSA math test topics,
- Over 2,500 practice problems covering all topics tested,
- The most important concepts you need to know,
- Clear and concise, easy-to-follow sections,
- Well designed for enhanced learning and interest,
- Hands-on experience with all question types,
- **2 full-length practice tests** with detailed answer explanations,
- Cost-Effective Pricing,

Powerful math exercises to help you avoid traps and pacing yourself to beat the FSA Florida test. Students will gain valuable experience and raise their confidence by taking 7th grade math practice tests, learning about test structure, and gaining a deeper understanding of what is tested on the FSA math grade 7. If ever there was a book to respond to the pressure to increase students' test scores, this is it.

WWW.MathNotion.COM

… So Much More Online!

- ✓ FREE Math Lessons

- ✓ More Math Learning Books!

- ✓ Mathematics Worksheets

- ✓ Online Math Tutors

For a PDF Version of This Book

Please Visit WWW.MathNotion.com

FSA Subject Test Mathematics Grade 7

Contents

Chapter 1 : Integers and Number Theory 11
  Rounding 12
  Rounding and Estimates 13
  Adding and Subtracting Integers 14
  Multiplying and Dividing Integers 15
  Order of Operations 16
  Ordering Integers and Numbers 17
  Integers and Absolute Value 18
  Factoring Numbers 19
  Greatest Common Factor 20
  Least Common Multiple 21
  Answers of Worksheets 22

Chapter 2 : Fractions 25
  Simplifying Fractions 26
  Adding and Subtracting Fractions 27
  Multiplying and Dividing Fractions 28
  Adding and Subtracting Mixed Numbers 29
  Multiplying and Dividing Mixed Numbers 30
  Answers of Worksheets 31

Chapter 3 : Decimals 33
  Adding and Subtracting Decimals 34
  Multiplying and Dividing Decimals 35
  Comparing Decimals 36
  Rounding Decimals 37
  Answers of Worksheets 38

Chapter 4 : Proportions, Ratios, and Percent 39
  Simplifying Ratios 40
  Proportional Ratios 41
  Similarity and Ratios 42
  Ratio and Rates Word Problems 43
  Percentage Calculations 44
  Percent Problems 45
  Discount, Tax and Tip 46

FSA Subject Test Mathematics Grade 7

Percent of Change...47
Simple Interest..48
Answers of Worksheets...49

Chapter 5 : Exponents and Radicals Expressions..52
Multiplication Property of Exponents...53
Zero and Negative Exponents..54
Division Property of Exponents..55
Powers of Products and Quotients..56
Negative Exponents and Negative Bases..57
Scientific Notation..58
Square Roots...59
Answers of Worksheets...60

Chapter 6 : Algebraic Expressions..63
Translate Phrases into an Algebraic Statement...64
Simplifying Variable Expressions...65
The Distributive Property...66
Evaluating One Variable Expressions..67
Evaluating Two Variables Expressions..68
Combining like Terms..69
Answers of Worksheets...70

Chapter 7 : Equations and Inequalities...72
One–Step Equations..73
Multi–Step Equations...74
Graphing Single-Variable Inequalities..75
One–Step Inequalities..76
Multi-Step Inequalities...77
Systems of Equations..78
Systems of Equations Word Problems..79
Answers of Worksheets...80

Chapter 8 : Linear Functions..83
Finding Slope...84
Graphing Lines Using Line Equation...85
Writing Linear Equations...86
Graphing Linear Inequalities..87
Finding Midpoint..88
Finding Distance of Two Points...89
Answers of Worksheets...90

WWW.MathNotion.Com

FSA Subject Test Mathematics Grade 7

Chapter 9 : Transformations 93
   Translations 94
   Reflections 95
   Rotations 97
   Dilations 99
   Coordinates of Vertices 100
   Answers of Worksheets 101

Chapter 10 : Geometry and Solid Figures 104
   Angles 105
   Pythagorean Relationship 106
   Triangles 107
   Polygons 108
   Trapezoids 109
   Circles 110
   Cubes 111
   Rectangular Prism 111
   Cylinder 113
   Pyramids and Cone 114
   Answers of Worksheets 115

Chapter 11 : Statistics and Probability 117
   Mean and Median 118
   Mode and Range 119
   Times Series 120
   Stem-and-Leaf Plot 121
   Pie Graph 122
   Probability Problems 123
   Answers of Worksheets 124

Chapter 12 : FSA Math Practice Tests 127
   FSA GRADE 7 MAHEMATICS REFRENCE MATERIALS 129
   FSA Practice Test 1 131
      Session 1 132
      Session 2 138
   FSA Practice Test 2 145
      Session 1 146
      Session 2 152

Chapter 13 : Answers and Explanations 159
   Answer Key 159

FSA Subject Test Mathematics Grade 7

Practice Test 1 161
Practice Test 2 167

FSA Subject Test Mathematics Grade 7

Chapter 1:
Integers and Number Theory

Topics that you will practice in this chapter:

- ✓ Rounding
- ✓ Rounding and Estimates
- ✓ Adding and Subtracting Integers
- ✓ Multiplying and Dividing Integers
- ✓ Order of Operations
- ✓ Ordering Integers and Numbers
- ✓ Integers and Absolute Value
- ✓ Factoring Numbers
- ✓ Greatest Common Factor (GCF)
- ✓ Least Common Multiple (LCM)

"Wherever there is number, there is beauty." −Proclus

FSA Subject Test Mathematics Grade 7

Rounding

✎ **Round each number to the nearest ten.**

1) 42 = ___   5) 19 = ___   9) 48 = ___

2) 88 = ___   6) 25 = ___   10) 81 = ___

3) 24 = ___   7) 93 = ___   11) 58 = ___

4) 57 = ___   8) 71 = ___   12) 87 = ___

✎ **Round each number to the nearest hundred.**

13) 198 = ___   17) 321 = ___   21) 580 = ___

14) 387 = ___   18) 433 = ___   22) 868 = ___

15) 816 = ___   19) 579 = ___   23) 480 = ___

16) 101 = ___   20) 825 = ___   24) 287 = ___

✎ **Round each number to the nearest thousand.**

25) 1,382 = ___   29) 9,099 = ___   33) 52,866 = ___

26) 3,420 = ___   30) 22,980 = ___   34) 85,190 = ___

27) 4,254 = ___   31) 45,188 = ___   35) 70,990 = ___

28) 6,861 = ___   32) 16,808 = ___   36) 26,869 = ___

WWW.MathNotion.Com

FSA Subject Test Mathematics Grade 7

Rounding and Estimates

✏ **Estimate the sum by rounding each number to the nearest ten.**

1) 13 + 22 = _____

2) 71 + 23 = _____

3) 61 + 58 = _____

4) 56 + 85 = _____

5) 368 + 249 = _____

6) 330 + 903 = _____

7) 471 + 293 = _____

8) 1,950 + 2,655 = _____

✏ **Estimate the product by rounding each number to the nearest ten.**

9) 32 × 71 = _____

10) 12 × 33 = _____

11) 31 × 83 = _____

12) 19 × 11 = _____

13) 42 × 76 = _____

14) 63 × 34 = _____

15) 19 × 31 = _____

16) 59 × 71 = _____

✏ **Estimate the sum or product by rounding each number to the nearest ten.**

17) $\begin{array}{r} 29 \\ \times\ 12 \\ \hline \end{array}$

18) $\begin{array}{r} 37 \\ \times\ 26 \\ \hline \end{array}$

19) $\begin{array}{r} 48 \\ +\ 82 \\ \hline \end{array}$

20) $\begin{array}{r} 65 \\ +44 \\ \hline \end{array}$

21) $\begin{array}{r} 37 \\ \times\ 14 \\ \hline \end{array}$

22) $\begin{array}{r} 71 \\ +\ 32 \\ \hline \end{array}$

WWW.MathNotion.Com

FSA Subject Test Mathematics Grade 7

Adding and Subtracting Integers

✍ **Find each sum.**

1) $14 + (-6) =$

2) $(-13) + (-20) =$

3) $5 + (-28) =$

4) $50 + (-12) =$

5) $(-7) + (-15) + 3 =$

6) $30 + (-14) + 8 =$

7) $40 + (-10) + (-14) + 17 =$

8) $(-15) + (-20) + 13 + 35 =$

9) $40 + (-20) + (38 - 29) =$

10) $28 + (-12) + (30 - 12) =$

✍ **Find each difference.**

11) $(-18) - (-7) =$

12) $25 - (-14) =$

13) $(-20) - 36 =$

14) $34 - (-19) =$

15) $51 - (30 - 21) =$

16) $17 - (5) - (-24) =$

17) $(35 + 20) - (-46) =$

18) $48 - 16 - (-8) =$

19) $62 - (28 + 17) - (-15) =$

20) $58 - (-23) - (-31) =$

21) $19 - (-8) - (-13) =$

22) $(19 - 24) - (-14) =$

23) $27 - 33 - (-21) =$

24) $58 - (32 + 24) - (-9) =$

25) $36 - (-30) + (-17) =$

26) $27 - (-42) + (-31) =$

WWW.MathNotion.Com

FSA Subject Test Mathematics Grade 7

Multiplying and Dividing Integers

✎ **Find each product.**

1) $(-9) \times (-5) =$

2) $(-3) \times 9 =$

3) $8 \times (-12) =$

4) $(-7) \times (-20) =$

5) $(-3) \times (-5) \times 6 =$

6) $(14 - 3) \times (-8) =$

7) $12 \times (-9) \times (-3) =$

8) $(140 + 10) \times (-2) =$

9) $10 \times (-12 + 8) \times 3 =$

10) $(-8) \times (-5) \times (-10) =$

✎ **Find each quotient.**

11) $42 \div (-7) =$

12) $(-48) \div (-6) =$

13) $(-40) \div (-8) =$

14) $54 \div (-2) =$

15) $152 \div 19 =$

16) $(-144) \div (-12) =$

17) $180 \div (-10) =$

18) $(-312) \div (-12) =$

19) $221 \div (-13) =$

20) $(-126) \div (6) =$

21) $(-161) \div (-7) =$

22) $-266 \div (-14) =$

23) $(-120) \div (-4) =$

24) $270 \div (-18) =$

25) $(-208) \div (-8) =$

26) $(135) \div (-15) =$

WWW.MathNotion.Com

FSA Subject Test Mathematics Grade 7

Order of Operations

✏️ **Evaluate each expression.**

1) $7 + (5 \times 4) =$

2) $14 - (3 \times 6) =$

3) $(19 \times 4) + 16 =$

4) $(16 - 7) - (8 \times 2) =$

5) $27 + (18 \div 3) =$

6) $(18 \times 8) \div 6 =$

7) $(32 \div 4) \times (-2) =$

8) $(9 \times 4) + (32 - 18) =$

9) $24 + (4 \times 3) + 7 =$

10) $(36 \times 3) \div (2 + 2) =$

11) $(-7) + (12 \times 3) + 11 =$

12) $(8 \times 5) - (24 \div 6) =$

13) $(7 \times 6 \div 3) - (12 + 9) =$

14) $(13 + 5 - 14) \times 3 - 2 =$

15) $(20 - 14 + 30) \times (64 \div 4) =$

16) $32 + (28 - (36 \div 9)) =$

17) $(7 + 6 - 4 - 7) + (15 \div 5) =$

18) $(85 - 20) + (20 - 18 + 7) =$

19) $(20 \times 2) + (14 \times 3) - 22 =$

20) $18 + 5 - (30 \times 3) + 20 =$

WWW.MathNotion.Com

FSA Subject Test Mathematics Grade 7

Ordering Integers and Numbers

✎ **Order each set of integers from least to greatest.**

1) $8, -10, -5, -3, 4$ ___, ___, ___, ___, ___, ___

2) $-10, -18, 6, 14, 27$ ___, ___, ___, ___, ___, ___

3) $15, -8, -21, 21, -23$ ___, ___, ___, ___, ___, ___

4) $-14, -40, 23, -12, 47$ ___, ___, ___, ___, ___, ___

5) $59, -54, 32, -57, 36$ ___, ___, ___, ___, ___, ___

6) $68, 26, -19, 47, -34$ ___, ___, ___, ___, ___, ___

✎ **Order each set of integers from greatest to least.**

7) $18, 36, -16, -18, -10$ ___, ___, ___, ___, ___, ___

8) $27, 34, -12, -24, 94$ ___, ___, ___, ___, ___, ___

9) $50, -21, -13, 42, -2$ ___, ___, ___, ___, ___, ___

10) $37, 46, -20, -16, 86$ ___, ___, ___, ___, ___, ___

11) $-18, 88, -26, -59, 75$ ___, ___, ___, ___, ___, ___

12) $-65, -30, -25, 3, 14$ ___, ___, ___, ___, ___, ___

WWW.MathNotion.Com

FSA Subject Test Mathematics Grade 7

Integers and Absolute Value

✎ **Write absolute value of each number.**

1) $|-2| =$

2) $|-27| =$

3) $|-20| =$

4) $|14| =$

5) $|6| =$

6) $|-55| =$

7) $|16| =$

8) $|2| =$

9) $|54| =$

10) $|-4| =$

11) $|-11|$

12) $|88| =$

13) $|0| =$

14) $|79| =$

15) $|-32| =$

16) $|-17| =$

17) $|42| =$

18) $|-46| =$

19) $|1| =$

20) $|-40| =$

✎ **Evaluate the value.**

21) $|-5| - \frac{|-21|}{7} =$

22) $14 - |3 - 15| - |-4| =$

23) $\frac{|-32|}{4} \times |-4| =$

24) $\frac{|7 \times (-3)|}{7} \times \frac{|-19|}{3} =$

25) $|4 \times (-5)| + \frac{|-40|}{5} =$

26) $\frac{|-45|}{9} \times \frac{|-24|}{12} =$

27) $|-12 + 8| \times \frac{|-7 \times 7|}{7} =$

28) $\frac{|-11 \times 2|}{4} \times |-16| =$

Factoring Numbers

✏ **List all positive factors of each number.**

1) 9

2) 16

3) 24

4) 30

5) 26

6) 46

7) 20

8) 68

9) 28

10) 98

11) 14

12) 54

13) 55

14) 18

15) 63

16) 34

17) 50

18) 62

19) 95

20) 64

21) 70

22) 45

23) 22

24) 65

FSA Subject Test Mathematics Grade 7

Greatest Common Factor

✏ Find the GCF for each number pair.

1) 6, 2

2) 4, 5

3) 3, 12

4) 7, 3

5) 5, 10

6) 8, 48

7) 6, 18

8) 9, 15

9) 12, 18

10) 4, 36

11) 6, 10

12) 28, 52

13) 25, 10

14) 22, 24

15) 9, 54

16) 8, 54

17) 42, 14

18) 16, 40

19) 9, 2, 3

20) 5, 15, 10

21) 7, 9, 2

22) 16, 64

23) 30, 48

24) 36, 63

WWW.MathNotion.Com

FSA Subject Test Mathematics Grade 7

Least Common Multiple

✏ **Find the LCM for each number pair.**

1) 6, 9

2) 15, 45

3) 16, 40

4) 12, 36

5) 18, 27

6) 14, 42

7) 6, 30

8) 8, 56

9) 7, 21

10) 8, 20

11) 15, 25

12) 7, 9

13) 4, 11

14) 8, 28

15) 28, 56

16) 40, 50

17) 12, 13

18) 22, 11

19) 36, 20

20) 15, 35

21) 18, 81

22) 30, 54

23) 18, 45

24) 75, 25

FSA Subject Test Mathematics Grade 7

Answers of Worksheets

Rounding

1) 40
2) 90
3) 20
4) 60
5) 20
6) 30
7) 90
8) 70
9) 50
10) 80
11) 60
12) 90
13) 200
14) 400
15) 800
16) 100
17) 300
18) 400
19) 600
20) 800
21) 600
22) 900
23) 500
24) 300
25) 1,000
26) 3,000
27) 4,000
28) 7,000
29) 9,000
30) 23,000
31) 45,000
32) 17,000
33) 53,000
34) 85,000
35) 71,000
36) 27,000

Rounding and Estimates

1) 30
2) 90
3) 120
4) 150
5) 620
6) 1,230
7) 760
8) 4,610
9) 2,100
10) 300
11) 2,400
12) 200
13) 3,200
14) 1,800
15) 600
16) 4,200
17) 300
18) 1,200
19) 130
20) 110
21) 400
22) 100

Adding and Subtracting Integers

1) 8
2) −33
3) −23
4) 38
5) −19
6) 24
7) 33
8) 13
9) 29
10) 34
11) −11
12) 39
13) −56
14) 53
15) 42
16) 36
17) 101
18) 40
19) 32
20) 112
21) 40
22) 9
23) 15
24) 11
25) 49
26) 38

Multiplying and Dividing Integers

1) 45
2) −27
3) −96
4) 140
5) 90
6) −88
7) 324
8) −300
9) −120
10) −400
11) −6
12) 8
13) 5
14) −27
15) 8
16) 12
17) −18
18) 26
19) −17
20) −21

FSA Subject Test Mathematics Grade 7

21) 23
22) 19
23) 30
24) −15
25) 26
26) −9

Order of Operations

1) 27
2) −4
3) 92
4) −7
5) 33
6) 24
7) −16
8) 50
9) 43
10) 27
11) 40
12) 36
13) −7
14) 10
15) 576
16) 56
17) 5
18) 74
19) 60
20) −47

Ordering Integers and Numbers

1) −10, −5, −3, 4, 8
2) −18, −10, 6, 14, 27
3) −23, −21, −8, 15, 21
4) −40, −14, −12, 23, 47
5) −57, −54, 32, 36, 59
6) −34, −19, 26, 47, 68
7) 36, 18, −10, −16, −18
8) 94, 34, 27, −12, −24
9) 50, 42, −2, −13, −21
10) 86, 46, 37, −16, −20
11) 88, 75, −18, −26, −59
12) 14, 3, −25, −30, −65

Integers and Absolute Value

1) 2
2) 27
3) 20
4) 14
5) 6
6) 55
7) 16
8) 2
9) 54
10) 4
11) 11
12) 88
13) 0
14) 79
15) 32
16) 17
17) 42
18) 46
19) 1
20) 40
21) 2
22) −2
23) 32
24) 19
25) 28
26) 10
27) 28
28) 88

Factoring Numbers

1) 1, 3, 9
2) 1, 2, 4, 8, 16
3) 1, 2, 3, 4, 6, 8, 12, 24
4) 1, 2, 3, 5, 6, 10, 15, 30
5) 1, 2, 13, 26
6) 1, 2, 23, 46
7) 1, 2, 4, 5, 10, 20
8) 1, 2, 4, 17, 34, 68
9) 1, 2, 4, 7, 14, 28
10) 1, 2, 7, 14, 49, 98
11) 1, 2, 7, 14
12) 1, 2, 3, 6, 9, 18, 27, 54
13) 1, 5, 11, 55
14) 1, 2, 3, 6, 9, 18
15) 1, 3, 7, 9, 21, 63
16) 1, 2, 17, 34
17) 1, 2, 5, 10, 25, 50
18) 1, 2, 31, 62
19) 1, 5, 19, 95
20) 1, 2, 4, 8, 16, 32, 64
21) 1, 2, 5, 7, 10, 14, 35, 70
22) 1, 3, 5, 9, 15, 45
23) 1, 2, 11, 22
24) 1, 5, 13, 65

Greatest Common Factor

1) 2
2) 1
3) 3
4) 1
5) 5
6) 8
7) 6
8) 3
9) 6
10) 4
11) 2
12) 4
13) 5
14) 2
15) 9
16) 2
17) 14
18) 8
19) 1
20) 5
21) 1
22) 16
23) 6
24) 9

Least Common Multiple

1) 18
2) 45
3) 80
4) 36
5) 54
6) 42
7) 30
8) 56
9) 21
10) 40
11) 75
12) 63
13) 44
14) 56
15) 56
16) 200
17) 156
18) 22
19) 180
20) 105
21) 162
22) 270
23) 90
24) 75

FSA Subject Test Mathematics Grade 7

Chapter 2 :
Fractions

Topics that you will practice in this chapter:

- ✓ Simplifying Fractions
- ✓ Adding and Subtracting Fractions
- ✓ Multiplying and Dividing Fractions
- ✓ Adding and Subtract Mixed Numbers
- ✓ Multiplying and Dividing Mixed Numbers

"A Man is like a fraction whose numerator is what he is and whose denominator is what he thinks of himself. The larger the denominator, the smaller the fraction." –Tolstoy

FSA Subject Test Mathematics Grade 7

Simplifying Fractions

✎ Simplify each fraction to its lowest terms.

1) $\dfrac{5}{10} =$

2) $\dfrac{28}{35} =$

3) $\dfrac{27}{36} =$

4) $\dfrac{40}{80} =$

5) $\dfrac{14}{56} =$

6) $\dfrac{32}{48} =$

7) $\dfrac{52}{65} =$

8) $\dfrac{15}{60} =$

9) $\dfrac{80}{160} =$

10) $\dfrac{55}{77} =$

11) $\dfrac{28}{112} =$

12) $\dfrac{32}{64} =$

13) $\dfrac{63}{72} =$

14) $\dfrac{81}{90} =$

15) $\dfrac{35}{105} =$

16) $\dfrac{25}{70} =$

17) $\dfrac{80}{280} =$

18) $\dfrac{12}{81} =$

19) $\dfrac{36}{186} =$

20) $\dfrac{240}{540} =$

21) $\dfrac{70}{560} =$

✎ Find the answer for each problem.

22) Which of the following fractions equal to $\dfrac{3}{4}$? ____

   A. $\dfrac{60}{90}$ B. $\dfrac{43}{104}$ C. $\dfrac{48}{64}$ D. $\dfrac{150}{300}$

23) Which of the following fractions equal to $\dfrac{5}{8}$? ____

   A. $\dfrac{125}{200}$ B. $\dfrac{115}{200}$ C. $\dfrac{50}{100}$ D. $\dfrac{30}{90}$

24) Which of the following fractions equal to $\dfrac{3}{7}$? ____

   A. $\dfrac{58}{116}$ B. $\dfrac{54}{126}$ C. $\dfrac{270}{167}$ D. $\dfrac{42}{63}$

WWW.MathNotion.Com

FSA Subject Test Mathematics Grade 7

Adding and Subtracting Fractions

✎ **Find the sum.**

1) $\frac{5}{9} + \frac{4}{9} =$

2) $\frac{1}{2} + \frac{1}{7} =$

3) $\frac{3}{8} + \frac{1}{4} =$

4) $\frac{3}{5} + \frac{1}{2} =$

5) $\frac{1}{4} + \frac{3}{5} =$

6) $\frac{7}{8} + \frac{3}{8} =$

7) $\frac{1}{2} + \frac{7}{10} =$

8) $\frac{2}{5} + \frac{2}{3} =$

9) $\frac{5}{7} + \frac{2}{3} =$

10) $\frac{7}{12} + \frac{3}{4} =$

11) $\frac{5}{6} + \frac{2}{5} =$

12) $\frac{1}{12} + \frac{2}{3} =$

✎ **Find the difference.**

13) $\frac{1}{3} - \frac{1}{6} =$

14) $\frac{3}{4} - \frac{1}{8} =$

15) $\frac{1}{2} - \frac{1}{3} =$

16) $\frac{1}{4} - \frac{1}{5} =$

17) $\frac{5}{8} - \frac{2}{3} =$

18) $\frac{1}{4} - \frac{1}{7} =$

19) $\frac{5}{6} - \frac{1}{9} =$

20) $\frac{3}{4} - \frac{1}{6} =$

21) $\frac{7}{8} - \frac{1}{12} =$

22) $\frac{8}{15} - \frac{3}{5} =$

23) $\frac{3}{12} - \frac{1}{14} =$

24) $\frac{10}{13} - \frac{7}{26} =$

25) $\frac{6}{7} - \frac{3}{4} =$

26) $\frac{4}{5} - \frac{1}{8} =$

27) $\frac{4}{7} - \frac{2}{35} =$

28) $\frac{9}{16} - \frac{2}{8} =$

29) $\frac{8}{9} - \frac{7}{18} =$

30) $\frac{1}{2} - \frac{4}{9} =$

WWW.MathNotion.Com

Multiplying and Dividing Fractions

✎ Find the value of each expression in lowest terms.

1) $\dfrac{1}{5} \times \dfrac{15}{5} =$

2) $\dfrac{9}{12} \times \dfrac{4}{9} =$

3) $\dfrac{1}{16} \times \dfrac{8}{10} =$

4) $\dfrac{1}{24} \times \dfrac{8}{10} =$

5) $\dfrac{1}{5} \times \dfrac{1}{4} =$

6) $\dfrac{7}{9} \times \dfrac{1}{7} =$

7) $\dfrac{6}{7} \times \dfrac{1}{3} =$

8) $\dfrac{2}{8} \times \dfrac{2}{8} =$

9) $\dfrac{5}{8} \times \dfrac{3}{5} =$

10) $\dfrac{4}{7} \times \dfrac{1}{8} =$

11) $\dfrac{7}{15} \times \dfrac{5}{7} =$

12) $\dfrac{3}{10} \times \dfrac{5}{9} =$

✎ Find the value of each expression in lowest terms.

13) $\dfrac{1}{4} \div \dfrac{1}{8} =$

14) $\dfrac{1}{10} \div \dfrac{1}{5} =$

15) $\dfrac{3}{4} \div \dfrac{1}{5} =$

16) $\dfrac{1}{3} \div \dfrac{5}{6} =$

17) $\dfrac{1}{7} \div \dfrac{8}{42} =$

18) $\dfrac{3}{4} \div \dfrac{1}{6} =$

19) $\dfrac{2}{7} \div \dfrac{7}{13} =$

20) $\dfrac{1}{24} \div \dfrac{3}{16} =$

21) $\dfrac{7}{12} \div \dfrac{5}{6} =$

22) $\dfrac{22}{18} \div \dfrac{11}{9} =$

23) $\dfrac{9}{35} \div \dfrac{3}{7} =$

24) $\dfrac{2}{7} \div \dfrac{8}{21} =$

25) $\dfrac{1}{9} \div \dfrac{2}{5} =$

26) $\dfrac{5}{12} \div \dfrac{3}{5} =$

27) $\dfrac{3}{20} \div \dfrac{1}{6} =$

28) $\dfrac{8}{20} \div \dfrac{3}{4} =$

29) $\dfrac{5}{6} \div \dfrac{2}{9} =$

30) $\dfrac{5}{11} \div \dfrac{3}{4} =$

WWW.MathNotion.Com

FSA Subject Test Mathematics Grade 7

Adding and Subtracting Mixed Numbers

✎ **Find the sum.**

1) $3\frac{1}{3} + 2\frac{1}{6} =$

2) $4\frac{1}{2} + 3\frac{1}{2} =$

3) $3\frac{3}{8} + 1\frac{1}{8} =$

4) $2\frac{1}{4} + 2\frac{1}{3} =$

5) $3\frac{5}{6} + 2\frac{7}{12} =$

6) $5\frac{4}{15} + 3\frac{3}{5} =$

7) $2\frac{1}{3} + 4\frac{3}{7} =$

8) $3\frac{1}{2} + 4\frac{2}{5} =$

9) $5\frac{2}{5} + 6\frac{3}{7} =$

10) $8\frac{5}{16} + 6\frac{1}{12} =$

✎ **Find the difference.**

11) $3\frac{1}{4} - 1\frac{3}{4} =$

12) $6\frac{3}{5} - 4\frac{2}{5} =$

13) $4\frac{1}{3} - 3\frac{1}{9} =$

14) $7\frac{1}{7} - 5\frac{1}{2} =$

15) $5\frac{1}{3} - 2\frac{1}{12} =$

16) $8\frac{1}{5} - 4\frac{1}{3} =$

17) $9\frac{1}{4} - 6\frac{1}{8} =$

18) $11\frac{7}{15} - 8\frac{3}{5} =$

19) $14\frac{5}{6} - 11\frac{3}{5} =$

20) $18\frac{2}{7} - 14\frac{1}{5} =$

21) $9\frac{1}{3} - 4\frac{1}{4} =$

22) $6\frac{1}{8} - 4\frac{1}{16} =$

23) $19\frac{3}{8} - 15\frac{1}{3} =$

24) $11\frac{1}{9} - 8\frac{1}{8} =$

25) $17\frac{1}{7} - 11\frac{1}{5} =$

26) $16\frac{2}{9} - 9\frac{5}{7} =$

WWW.MathNotion.Com

FSA Subject Test Mathematics Grade 7

Multiplying and Dividing Mixed Numbers

✏️ **Find the product.**

1) $5\frac{1}{2} \times 2\frac{1}{4} =$

2) $5\frac{1}{3} \times 4\frac{1}{3} =$

3) $5\frac{3}{4} \times 6\frac{1}{4} =$

4) $3\frac{1}{3} \times 2\frac{3}{5} =$

5) $4\frac{8}{10} \times 1\frac{1}{24} =$

6) $6\frac{2}{7} \times 1\frac{1}{11} =$

7) $8\frac{2}{3} \times 3\frac{1}{2} =$

8) $3\frac{4}{7} \times 2\frac{1}{5} =$

9) $5\frac{2}{8} \times 4\frac{1}{6} =$

10) $7\frac{3}{3} \times 1\frac{3}{8} =$

✏️ **Find the quotient.**

11) $2\frac{2}{5} \div 4\frac{1}{5} =$

12) $4\frac{1}{6} \div 3\frac{1}{3} =$

13) $6\frac{1}{3} \div 1\frac{1}{2} =$

14) $7\frac{1}{10} \div 2\frac{2}{5} =$

15) $3\frac{1}{3} \div 1\frac{1}{9} =$

16) $1\frac{1}{10} \div 4\frac{1}{2} =$

17) $1\frac{3}{16} \div 5\frac{1}{4} =$

18) $4\frac{1}{3} \div 4\frac{3}{4} =$

19) $9\frac{1}{3} \div 2\frac{1}{4} =$

20) $15\frac{1}{3} \div 5\frac{1}{2} =$

21) $4\frac{1}{6} \div 1\frac{1}{5} =$

22) $1\frac{1}{18} \div 1\frac{2}{9} =$

23) $4\frac{2}{7} \div 1\frac{3}{10} =$

24) $7\frac{1}{3} \div 2\frac{2}{11} =$

25) $8\frac{2}{5} \div 1\frac{1}{6} =$

26) $9\frac{1}{3} \div 2\frac{1}{7} =$

WWW.MathNotion.Com

FSA Subject Test Mathematics Grade 7

Answers of Worksheets

Simplifying Fractions

1) $\frac{1}{2}$ 7) $\frac{4}{5}$ 13) $\frac{7}{8}$ 19) $\frac{6}{31}$

2) $\frac{4}{5}$ 8) $\frac{1}{4}$ 14) $\frac{9}{10}$ 20) $\frac{4}{9}$

3) $\frac{3}{4}$ 9) $\frac{1}{2}$ 15) $\frac{1}{3}$ 21) $\frac{1}{8}$

4) $\frac{1}{2}$ 10) $\frac{5}{7}$ 16) $\frac{5}{14}$ 22) C

5) $\frac{1}{4}$ 11) $\frac{1}{4}$ 17) $\frac{2}{7}$ 23) A

6) $\frac{2}{3}$ 12) $\frac{1}{2}$ 18) $\frac{4}{27}$ 24) B

Adding and Subtracting Fractions

1) $\frac{9}{9}=1$ 9) $1\frac{8}{21}$ 17) $-\frac{1}{24}$ 25) $\frac{3}{28}$

2) $\frac{9}{14}$ 10) $1\frac{1}{3}$ 18) $\frac{3}{28}$ 26) $\frac{27}{40}$

3) $\frac{5}{8}$ 11) $1\frac{7}{30}$ 19) $\frac{13}{18}$ 27) $\frac{18}{35}$

4) $1\frac{1}{10}$ 12) $\frac{3}{4}$ 20) $\frac{7}{12}$ 28) $\frac{5}{16}$

5) $\frac{17}{20}$ 13) $\frac{1}{6}$ 21) $\frac{19}{24}$ 29) $\frac{1}{2}$

6) $1\frac{1}{4}$ 14) $\frac{5}{8}$ 22) $-\frac{1}{15}$ 30) $\frac{1}{18}$

7) $1\frac{1}{5}$ 15) $\frac{1}{6}$ 23) $\frac{5}{28}$

8) $1\frac{1}{15}$ 16) $\frac{1}{20}$ 24) $\frac{1}{2}$

Multiplying and Dividing Fractions

1) $\frac{3}{5}$ 5) $\frac{1}{20}$ 9) $\frac{3}{8}$ 13) 2

2) $\frac{1}{3}$ 6) $\frac{1}{9}$ 10) $\frac{1}{14}$ 14) $\frac{1}{2}$

3) $\frac{1}{20}$ 7) $\frac{2}{7}$ 11) $\frac{1}{3}$ 15) $3\frac{3}{4}$

4) $\frac{1}{30}$ 8) $\frac{1}{16}$ 12) $\frac{1}{6}$ 16) $\frac{2}{5}$

WWW.MathNotion.Com

FSA Subject Test Mathematics Grade 7

17) $\frac{3}{4}$
18) $4\frac{1}{2}$
19) $\frac{26}{49}$
20) $\frac{2}{9}$

21) $\frac{7}{10}$
22) 1
23) $\frac{3}{5}$
24) $\frac{3}{4}$

25) $\frac{5}{18}$
26) $\frac{25}{36}$
27) $\frac{9}{10}$
28) $\frac{8}{15}$

29) $3\frac{3}{4}$
30) $\frac{20}{33}$

Adding and Subtracting Mixed Numbers

1) $5\frac{1}{2}$
2) 8
3) $4\frac{1}{2}$
4) $4\frac{7}{12}$
5) $6\frac{5}{12}$
6) $8\frac{13}{15}$
7) $6\frac{16}{21}$

8) $7\frac{9}{10}$
9) $11\frac{29}{35}$
10) $14\frac{19}{48}$
11) $1\frac{1}{2}$
12) $2\frac{1}{5}$
13) $1\frac{2}{9}$
14) $1\frac{9}{14}$

15) $3\frac{1}{4}$
16) $3\frac{13}{15}$
17) $3\frac{1}{8}$
18) $2\frac{13}{15}$
19) $3\frac{7}{30}$
20) $4\frac{3}{35}$
21) $5\frac{1}{12}$

22) $2\frac{1}{16}$
23) $4\frac{1}{24}$
24) $2\frac{71}{72}$
25) $5\frac{33}{35}$
26) $6\frac{32}{63}$

Multiplying and Dividing Mixed Numbers

1) $12\frac{3}{8}$
2) $23\frac{1}{9}$
3) $35\frac{15}{16}$
4) $8\frac{2}{3}$
5) 5
6) $6\frac{6}{7}$
7) $30\frac{1}{3}$
8) $7\frac{6}{7}$
9) $21\frac{7}{8}$

10) 11
11) $\frac{4}{7}$
12) $1\frac{1}{4}$
13) $4\frac{2}{9}$
14) $2\frac{23}{24}$
15) 3
16) $\frac{11}{45}$
17) $\frac{19}{84}$
18) $\frac{52}{57}$

19) $4\frac{4}{27}$
20) $2\frac{26}{33}$
21) $3\frac{17}{36}$
22) $\frac{19}{22}$
23) $3\frac{27}{91}$
24) $3\frac{13}{36}$
25) $7\frac{1}{5}$
26) $4\frac{16}{45}$

FSA Subject Test Mathematics Grade 7

Chapter 3:
Decimals

Topics that you will practice in this chapter:

- ✓ Adding and Subtracting Decimals
- ✓ Multiplying and Dividing Decimals
- ✓ Comparing Decimals
- ✓ Rounding Decimals

*"The study of mathematics, like the Nile, begins in minuteness but ends in magnificence." –
Charles Caleb Colton*

FSA Subject Test Mathematics Grade 7

Adding and Subtracting Decimals

✏ **Add and subtract decimals.**

1) 35.19 − 24.28 = _____

2) 34.29 + 42.58 = _____

3) 61.20 + 33.75 = _____

4) 38.72 − 21.68 = _____

5) 57.39 + 26.54 = _____

6) 70.24 − 42.35 = _____

7) 86.09 − 35.14 = _____

8) 54.51 + 32.66 = _____

9) 114.21 − 88.69 = _____

✏ **Find the missing number.**

10) ___ + 2.8 = 5.4

11) 4.1 + ___ = 5.88

12) 6.45 + ___ = 8

13) 7.25 − ___ = 3.40

14) ___ − 2.35 = 4.25

15) ___ − 19.85 = 6.54

16) 22.15 + ___ = 28.95

17) ___ − 37.16 = 9.42

18) ___ + 24.50 = 34.19

19) 72.40 + ___ = 125.20

WWW.MathNotion.Com

FSA Subject Test Mathematics Grade 7

Multiplying and Dividing Decimals

✎ **Find the product.**

1) $0.5 \times 0.6 =$

2) $3.3 \times 0.4 =$

3) $1.28 \times 0.5 =$

4) $0.35 \times 0.6 =$

5) $1.85 \times 0.6 =$

6) $0.24 \times 0.5 =$

7) $5.25 \times 1.4 =$

8) $18.5 \times 4.6 =$

9) $15.4 \times 6.8 =$

10) $19.5 \times 2.6 =$

11) $32.2 \times 1.5 =$

12) $78.4 \times 4.5 =$

✎ **Find the quotient.**

13) $1.85 \div 10 =$

14) $74.6 \div 100 =$

15) $3.6 \div 3 =$

16) $9.6 \div 0.4 =$

17) $15.5 \div 0.5 =$

18) $32.8 \div 0.2 =$

19) $22.15 \div 1,000 =$

20) $53.55 \div 0.7 =$

21) $322.2 \div 0.2 =$

22) $50.67 \div 0.18 =$

23) $77.4 \div 0.8 =$

24) $27.93 \div 0.03 =$

FSA Subject Test Mathematics Grade 7

Comparing Decimals

✎ Write the correct comparison symbol (>, < or =).

1) 0.70 ☐ 0.070

2) 0.049 ☐ 0.49

3) 5.090 ☐ 5.09

4) 2.57 ☐ 2.05

5) 9.03 ☐ 0.930

6) 6.06 ☐ 6.6

7) 7.02 ☐ 7.020

8) 3.04 ☐ 3.2

9) 3.61 ☐ 3.245

10) 0.986 ☐ 0.0986

11) 17.24 ☐ 17.240

12) 0.759 ☐ 0.81

13) 9.040 ☐ 9.40

14) 5.73 ☐ 5.213

15) 9.44 ☐ 9.404

16) 7.17 ☐ 7.170

17) 4.85 ☐ 4.085

18) 9.041 ☐ 9.40

19) 3.033 ☐ 3.030

20) 4.97 ☐ 4.970

WWW.MathNotion.Com

FSA Subject Test Mathematics Grade 7

Rounding Decimals

✎ **Round each decimal to the nearest whole number.**

1) 28.12   3) 16.22   5) 7.95

2) 6.9   4) 8.5   6) 52.7

✎ **Round each decimal to the nearest tenth.**

7) 31.761   9) 94.729   11) 13.219

8) 14.421   10) 77.89   12) 59.89

✎ **Round each decimal to the nearest hundredth.**

13) 8.428   15) 55.3786   17) 62.241

14) 23.812   16) 231.912   18) 19.447

✎ **Round each decimal to the nearest thousandth.**

19) 15.54324   21) 243.8652   23) 67.1983

20) 34.62586   22) 80.4529   24) 72.36788

FSA Subject Test Mathematics Grade 7

Answers of Worksheets

Adding and Subtracting Decimals

1) 10.91
2) 76.87
3) 94.95
4) 17.04
5) 83.93
6) 27.89
7) 50.95
8) 87.17
9) 25.52
10) 2.6
11) 1.78
12) 1.55
13) 3.85
14) 6.6
15) 26.39
16) 6.8
17) 46.58
18) 9.69
19) 52.8

Multiplying and Dividing Decimals

1) 0.3
2) 1.32
3) 0.64
4) 0.21
5) 1.11
6) 0.12
7) 7.35
8) 85.1
9) 104.72
10) 50.7
11) 48.3
12) 352.8
13) 0.185
14) 0.746
15) 1.2
16) 24
17) 31
18) 164
19) 0.02215
20) 76.5
21) 1,611
22) 281.5
23) 96.75
24) 931

Comparing Decimals

1) >
2) <
3) =
4) >
5) >
6) <
7) =
8) <
9) >
10) >
11) =
12) <
13) <
14) >
15) >
16) =
17) >
18) <
19) >
20) =

Rounding Decimals

1) 28
2) 7
3) 16
4) 9
5) 8
6) 53
7) 31.8
8) 14.4
9) 94.7
10) 77.9
11) 13.2
12) 59.9
13) 8.43
14) 23.81
15) 55.38
16) 231.91
17) 62.24
18) 19.45
19) 15.543
20) 34.626
21) 243.865
22) 80.453
23) 67.198
24) 72.368

WWW.MathNotion.Com

FSA Subject Test Mathematics Grade 7

Chapter 4:
Proportions, Ratios, and Percent

Topics that you will practice in this chapter:

- ✓ Simplifying Ratios
- ✓ Proportional Ratios
- ✓ Similarity and Ratios
- ✓ Ratio and Rates Word Problems
- ✓ Percentage Calculations
- ✓ Percent Problems
- ✓ Discount, Tax and Tip
- ✓ Percent of Change
- ✓ Simple Interest

Without mathematics, there's nothing you can do. Everything around you is mathematics. Everything around you is numbers." – Shakuntala Devi

FSA Subject Test Mathematics Grade 7

Simplifying Ratios

✎ **Reduce each ratio.**

1) 15: 20 = ___: ___
2) 7: 70 = ___: ___
3) 16: 28 = ___: ___
4) 7: 21 = ___: ___
5) 4: 40 = ___: ___
6) 6: 48 = ___: ___
7) 16: 64 = ___: ___
8) 10: 25 = ___: ___

9) 8: 48 = ___: ___
10) 49: 63 = ___: ___
11) 18: 27 = ___: ___
12) 35: 10 = ___: ___
13) 90: 9 = ___: ___
14) 24: 32 = ___: ___
15) 7: 56 = ___: ___
16) 45: 63 = ___: ___

17) 56: 72 = ___: ___
18) 26: 13 = ___: ___
19) 15: 45 = ___: ___
20) 28: 4 = ___: ___
21) 24: 48 = ___: ___
22) 30: 24 = ___: ___
23) 70: 140 = ___: ___
24) 6: 180 = ___: ___

✎ **Write each ratio as a fraction in simplest form.**

25) 6: 12 =
26) 30: 50 =
27) 15: 35 =
28) 9: 27 =
29) 8: 24 =
30) 18: 84 =
31) 7: 14 =

32) 7: 35 =
33) 40: 96 =
34) 12: 54 =
35) 44: 52 =
36) 12: 27 =
37) 15: 180 =
38) 39: 143 =

39) 20: 300 =
40) 30: 120 =
41) 56: 42 =
42) 26: 130 =
43) 66: 123 =
44) 70: 630 =
45) 75: 125 =

WWW.MathNotion.Com

FSA Subject Test Mathematics Grade 7

Proportional Ratios

✎ **Fill in the blanks; Calculate each proportion.**

1) $3:8 = \_\_ : 48$

2) $2:5 = 20:\_\_$

3) $1:9 = \_\_ : 81$

4) $6:7 = 12:\_\_$

5) $9:2 = 63:\_\_$

6) $8:7 = \_\_ : 49$

7) $20:3 = \_\_ : 15$

8) $1:3 = \_\_ : 75$

9) $7:6 = \_\_ : 60$

10) $8:5 = \_\_ : 45$

11) $3:10 = 60:\_\_$

12) $6:11 = 42:\_\_$

✎ **State if each pair of ratios form a proportion.**

13) $\frac{3}{20}$ and $\frac{9}{60}$

14) $\frac{1}{7}$ and $\frac{6}{42}$

15) $\frac{3}{7}$ and $\frac{24}{56}$

16) $\frac{4}{9}$ and $\frac{12}{18}$

17) $\frac{1}{9}$ and $\frac{12}{81}$

18) $\frac{7}{8}$ and $\frac{21}{28}$

19) $\frac{9}{13}$ and $\frac{27}{39}$

20) $\frac{1}{8}$ and $\frac{8}{64}$

21) $\frac{6}{19}$ and $\frac{30}{85}$

22) $\frac{5}{9}$ and $\frac{40}{81}$

23) $\frac{9}{14}$ and $\frac{108}{168}$

24) $\frac{15}{23}$ and $\frac{360}{552}$

✎ **Calculate each proportion.**

25) $\frac{20}{25} = \frac{32}{x}$, $x = \_\_\_$

26) $\frac{1}{8} = \frac{32}{x}$, $x = \_\_\_$

27) $\frac{15}{5} = \frac{21}{x}$, $x = \_\_\_$

28) $\frac{1}{7} = \frac{x}{294}$, $x = \_\_\_$

29) $\frac{7}{9} = \frac{x}{81}$, $x = \_\_\_$

30) $\frac{1}{5} = \frac{13}{x}$, $x = \_\_\_$

31) $\frac{9}{5} = \frac{36}{x}$, $x = \_\_\_$

32) $\frac{6}{13} = \frac{48}{x}$, $x = \_\_\_$

33) $\frac{5}{8} = \frac{x}{88}$, $x = \_\_\_$

34) $\frac{4}{15} = \frac{x}{240}$, $x = \_\_\_$

35) $\frac{9}{19} = \frac{x}{266}$, $x = \_\_\_$

36) $\frac{7}{15} = \frac{x}{270}$, $x = \_\_\_$

WWW.MathNotion.Com

FSA Subject Test Mathematics Grade 7

Similarity and Ratios

✏️ **Each pair of figures is similar. Find the missing side.**

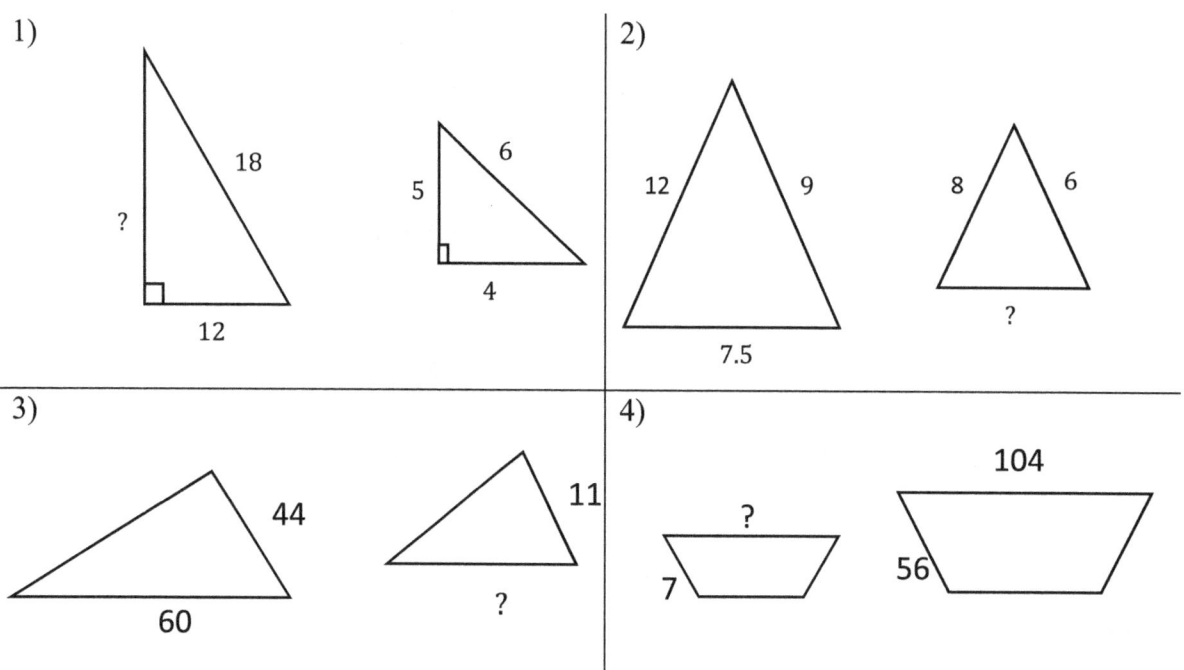

✏️ **Calculate.**

5) Two rectangles are similar. The first is 24 feet wide and 120 feet long. The second is 30 feet wide. What is the length of the second rectangle? _____

6) Two rectangles are similar. One is 5 meters by 36 meters. The longer side of the second rectangle is 90 meters. What is the other side of the second rectangle? _____

7) A building casts a shadow 25 ft long. At the same time a girl 10 ft tall casts a shadow 5 ft long. How tall is the building? _____

8) The scale of a map of Texas is 4 inches: 32 miles. If you measure the distance from Dallas to Martin County as 38.4 inches, approximately how far is Martin County from Dallas? _____

Ratio and Rates Word Problems

✏ **Find the answer for each word problem.**

1) Mason has 24 red cards and 36 green cards. What is the ratio of Mason's red cards to his green cards? _____

2) In a party, 45 soft drinks are required for every 54 guests. If there are 378 guests, how many soft drinks is required? _____

3) In Mason's class, 42 of the students are tall and 24 are short. In Michael's class 84 students are tall and 48 students are short. Which class has a higher ratio of tall to short students? _____

4) The price of 5 apples at the Quick Market is $4.6. The price of 7 of the same apples at Walmart is $5.95. Which place is the better buy? _____

5) The bakers at a Bakery can make 90 bagels in 3 hours. How many bagels can they bake in 24 hours? What is that rate per hour? _____

6) You can buy 5 cans of green beans at a supermarket for $5.75. How much does it cost to buy 45 cans of green beans? _____

7) The ratio of boys to girls in a class is 4: 7. If there are 32 boys in the class, how many girls are in that class? _____

8) The ratio of red marbles to blue marbles in a bag is 3: 7. If there are 50 marbles in the bag, how many of the marbles are red? _____

FSA Subject Test Mathematics Grade 7

Percentage Calculations

✏ **Calculate the given percent of each value.**

1) 3% of 60 = ___
2) 20% of 32 = ___
3) 4% of 72 = ___
4) 16% of 32 = ___
5) 25% of 124 = ___
6) 35% of 56 = ___

7) 15% of 20 = ___
8) 14% of 150 = ___
9) 80% of 50 = ___
10) 12% of 115 = ___
11) 72% of 250 = ___
12) 52% of 500 = ___

13) 70% of 400 = ___
14) 27% of 145 = ___
15) 90% of 64 = ___
16) 60% of 55 = ___
17) 22% of 210 = ___
18) 8% of 235 = ___

✏ **Calculate the percent of each given value.**

19) ___% of 25 = 5
20) ___% of 40 = 20
21) ___% of 25 = 2
22) ___% of 50 = 16
23) ___% of 250 = 5

24) ___% of 40 = 32
25) ___% of 125 = 20
26) ___% of 700 = 49
27) ___% of 350 = 49
28) ___% of 500 = 210

✏ **Calculate each percent problem.**

29) A Cinema has 250 seats. 60 seats were sold for the current movie. What percent of seats are empty? ___ %

30) There are 68 boys and 92 girls in a class. 75% of the students in the class take the bus to school. How many students do not take the bus to school? ___

WWW.MathNotion.Com

FSA Subject Test Mathematics Grade 7

Percent Problems

✎ **Calculate each problem.**

1) 9 is what percent of 45? ____%

2) 60 is what percent of 120? ____%

3) 10 is what percent of 200? ____%

4) 15 is what percent of 125? ____%

5) 10 is what percent of 400? ____%

6) 66 is what percent of 55? ____%

7) 40 is what percent of 160? ____%

8) 40 is what percent of 50? ____%

9) 120 is what percent of 800? ____%

10) 78 is what percent of 120? ____%

11) 36 is what percent of 144? ____%

12) 17 is what percent of 85? ____%

13) 90 is what percent of 900? ____%

14) 36 is what percent of 16? ____%

15) 63 is what percent of 14? ____%

16) 18 is what percent of 60? ____%

17) 126 is what percent of 200? ____%

18) 232 is what percent of 40? ____%

✎ **Calculate each percent word problem.**

19) There are 40 employees in a company. On a certain day, 25 were present. What percent showed up for work? ____%

20) A metal bar weighs 60 ounces. 25% of the bar is gold. How many ounces of gold are in the bar? _____

21) A crew is made up of 12 women; the rest are men. If 15% of the crew are women, how many people are in the crew? _____

22) There are 40 students in a class and 8 of them are girls. What percent are boys? ____%

23) The Royals softball team played 400 games and won 280 of them. What percent of the games did they lose? ____%

WWW.MathNotion.Com

FSA Subject Test Mathematics Grade 7

Discount, Tax and Tip

✏️ **Find the selling price of each item.**

1) Original price of a computer: $420
   Tax: 8% Selling price: $_____

2) Original price of a laptop: $280
   Tax: 4% Selling price: $_____

3) Original price of a sofa: $820
   Tax: 5% Selling price: $_____

4) Original price of a car: $15,800
   Tax: 3.6%   Selling price: $_____

5) Original price of a Table: $250
   Tax: 9% Selling price: $_____

6) Original price of a house: $630,000
   Tax: 1.8%   Selling price: $_____

7) Original price of a tablet: $450
   Discount: 30%   Selling price: $____

8) Original price of a chair: $390
   Discount: 8%   Selling price: $____

9) Original price of a book: $75
   Discount: 42%   Selling price: $____

10) Original price of a cellphone: $820
    Discount: 23%   Selling price: $___

11) Food bill: $45
    Tip: 15%   Price: $_____

12) Food bill: $32
    Tipp: 20%   Price: $_____

13) Food bill: $90
    Tip: 35%   Price: $_____

14) Food bill: $42
    Tipp: 12%   Price: $_____

✏️ **Find the answer for each word problem.**

15) Nicolas hired a moving company. The company charged $500 for its services, and Nicolas gives the movers a 40% tip. How much does Nicolas tip the movers? $_____

16) Mason has lunch at a restaurant and the cost of his meal is $90. Mason wants to leave a 25% tip. What is Mason's total bill including tip? $_____

17) The sales tax in Texas is 19.80% and an item costs $350. How much is the tax? $_____

18) The price of a table at Best Buy is $680. If the sales tax is 5%, what is the final price of the table including tax? $_____

WWW.MathNotion.Com

FSA Subject Test Mathematics Grade 7

Percent of Change

✎ **Find each percent of change.**

1) From 150 to 450. ___ %

2) From 50 ft to 250 ft. ___ %

3) From $60 to $360. ___ %

4) From 60 cm to 180 cm. ___ %

5) From 15 to 45. ___ %

6) From 80 to 16. ___ %

7) From 120 to 360. ___ %

8) From 900 to 450. ___ %

9) From 1,000 to 200. ___ %

10) From 144 to 36. ___ %

✎ **Calculate each percent of change word problem.**

11) Bob got a raise, and his hourly wage increased from $42 to $63. What is the percent increase? ___ %

12) The price of a pair of shoes increases from $50 to $61. What is the percent increase? ___ %

13) At a coffee shop, the price of a cup of coffee increased from $4.80 to $5.76. What is the percent increase in the cost of the coffee? ___ %

14) 51 cm are cut from 85 cm board. What is the percent decrease in length? ___ %

15) In a class, the number of students has been increased from 54 to 81. What is the percent increase? ___ %

16) The price of gasoline rises from $24.40 to $30.50 in one month. By what percent did the gas price rise? ___ %

17) A shirt was originally priced at $38. It went on sale for $24.70. What was the percent that the shirt was discounted? ___ %

WWW.MathNotion.Com

FSA Subject Test Mathematics Grade 7

Simple Interest

✍ **Determine the simple interest for these loans.**

1) $480 at 11% for 3 years. $ _____

2) $4,200 at 7% for 4 years. $ _____

3) $2,500 at 20% for 3 years. $ _____

4) $6,800 at 3.9% for 4 months. $ ____

5) $800 at 6% for 7 months. $ _____

6) $36,000 at 4.2% for 6 years. $ _____

7) $6,500 at 7% for 4 years. $ _____

8) $850 at 9.5% for 2 years. $ _____

9) $1,200 at 5.8% for 9 months. $ ___

10) $3,000 at 4.5% for 7 years. $ _____

✍ **Calculate each simple interest word problem.**

11) A new car, valued at $22,000, depreciates at 8.5% per year. What is the value of the car one year after purchase? $_____

12) Sara puts $9,000 into an investment yielding 6% annual simple interest; she left the money in for three years. How much interest does Sara get at the end of those three years? $_____

13) A bank is offering 12% simple interest on a savings account. If you deposit $16,400, how much interest will you earn in two years? $_____

14) $720 interest is earned on a principal of $6,000 at a simple interest rate of 4% interest per year. For how many years was the principal invested? _____

15) In how many years will $2,200 yield an interest of $440 at 4% simple interest? _____

16) Jim invested $8,000 in a bond at a yearly rate of 4.5%. He earned $1,440 in interest. How long was the money invested? _____

WWW.MathNotion.Com

FSA Subject Test Mathematics Grade 7

Answers of Worksheets

Simplifying Ratios

1) 3 : 4
2) 1 : 10
3) 4 : 7
4) 1 : 3
5) 1 : 10
6) 1 : 8
7) 2 : 8
8) 2 : 5
9) 1 : 6
10) 7 : 9
11) 2 : 3
12) 7 : 2
13) 10 : 1
14) 3 : 4
15) 1 : 8
16) 5 : 7
17) 7 : 9
18) 2 : 1
19) 1 : 3
20) 7 : 1
21) 1 : 2
22) 5 : 4
23) 1 : 2
24) 1 : 30
25) $\frac{1}{2}$
26) $\frac{3}{5}$
27) $\frac{3}{7}$
28) $\frac{1}{3}$
29) $\frac{1}{3}$
30) $\frac{3}{14}$
31) $\frac{1}{2}$
32) $\frac{1}{5}$
33) $\frac{5}{12}$
34) $\frac{2}{9}$
35) $\frac{11}{13}$
36) $\frac{4}{9}$
37) $\frac{1}{12}$
38) $\frac{3}{11}$
39) $\frac{1}{15}$
40) $\frac{1}{4}$
41) $\frac{4}{3}$
42) $\frac{1}{5}$
43) $\frac{22}{41}$
44) $\frac{1}{9}$
45) $\frac{3}{5}$

Proportional Ratios

1) 18
2) 50
3) 9
4) 14
5) 14
6) 56
7) 100
8) 25
9) 70
10) 72
11) 200
12) 77
13) Yes
14) Yes
15) Yes
16) No
17) No
18) No
19) Yes
20) Yes
21) No
22) No
23) Yes
24) Yes
25) 40
26) 256
27) 7
28) 42
29) 63
30) 65
31) 20
32) 104
33) 55
34) 64
35) 126
36) 126

Similarity and ratios

1) 15
2) 5
3) 15
4) 13
5) 150 feet
6) 12.5 meters
7) 50 feet
8) 307.2 miles

Ratio and Rates Word Problems

1) 2 : 3
2) 315

WWW.MathNotion.Com

FSA Subject Test Mathematics Grade 7

3) The ratio for both classes is 7 to 4.
4) Walmart is a better buy.
5) 720, the rate is 30 per hour.
6) $51.75
7) 56
8) 15

Percentage Calculations

1) 1.8
2) 6.4
3) 2.88
4) 5.12
5) 31
6) 19.6
7) 3
8) 21
9) 40
10) 13.8
11) 180
12) 260
13) 280
14) 39.15
15) 57.6
16) 33
17) 46.2
18) 18.8
19) 20%
20) 50%
21) 8%
22) 32%
23) 2%
24) 80%
25) 16%
26) 7%
27) 14%
28) 42%
29) 76%
30) 40

Percent Problems

1) 20%
2) 50%
3) 5%
4) 12%
5) 2.5%
6) 120%
7) 25%
8) 80%
9) 15%
10) 65%
11) 25%
12) 20%
13) 10%
14) 225%
15) 450%
16) 30%
17) 63%
18) 580%
19) 62.5%
20) 15 ounces
21) 80
22) 80%
23) 30%

Discount, Tax and Tip

1) $453.60
2) $291.20
3) $861.00
4) $16,368.80
5) $272.50
6) $641,340
7) $315.00
8) $358.80
9) $43.50
10) $631.40
11) $51.75
12) $38.40
13) $121.50
14) $47.04
15) $200.00
16) $112.50
17) $69.30
18) $714.00

FSA Subject Test Mathematics Grade 7

Percent of Change

1) 200%
2) 400%
3) 500%
4) 200%
5) 200%
6) 80%
7) 200%
8) 50%
9) 80%
10) 75%
11) 50%
12) 22%
13) 20%
14) 60%
15) 50%
16) 25%
17) 35%

Simple Interest

1) $158.40
2) $1,176.00
3) $1,500.00
4) $88.40
5) $28.00
6) $9,072.00
7) $1,820.00
8) $161.50
9) $52.20
10) $945.00
11) $20,130.00
12) $1,620.00
13) $3,936.00
14) 3 years
15) 5 years
16) 4 years

FSA Subject Test Mathematics Grade 7

Chapter 5 :

Exponents and Radicals Expressions

Topics that you will practice in this chapter:

- ✓ Multiplication Property of Exponents
- ✓ Zero and Negative Exponents
- ✓ Division Property of Exponents
- ✓ Powers of Products and Quotients
- ✓ Negative Exponents and Negative Bases
- ✓ Scientific Notation
- ✓ Square Roots

Mathematics is no more computation than typing is literature.

- John Allen Paulos

FSA Subject Test Mathematics Grade 7

Multiplication Property of Exponents

✎ Simplify and write the answer in exponential form.

1) $4 \times 4^5 =$

2) $8^4 \times 8 =$

3) $7^3 \times 7^3 =$

4) $9^2 \times 9^2 =$

5) $2^2 \times 2^4 \times 2 =$

6) $5 \times 5^3 \times 5^3 =$

7) $4^3 \times 4^2 \times 4 \times 4 =$

8) $5x \times x =$

9) $x^3 \times x^3 =$

10) $x^7 \times x^2 =$

11) $x^4 \times x^3 \times x^2 =$

12) $10x \times 3x =$

13) $4x^3 \times 4x^3 =$

14) $7x^3 \times x =$

15) $3x^2 \times 4x^2 \times x^2 =$

16) $5x^4 \times x^4 =$

17) $2x^8 \times 2x =$

18) $6x \times x^5 =$

19) $4x^2 \times 6x^6 =$

20) $5yx^3 \times 4x =$

21) $7x^3 \times y^5 x^7 =$

22) $y^2 x^3 \times y^5 x^4 =$

23) $3x^5 \times 4x^3 y^4 =$

24) $4x^4 \times 9x^2 y^5 =$

25) $5x^3 y^4 \times 6x^8 y^2 =$

26) $8x^3 y^6 \times 4xy^3 =$

27) $2xy^5 \times 6x^3 y^3 =$

28) $4x^5 y^2 \times 4x^2 y^8 =$

29) $7x \times 3y^8 x^2 \times y^5 =$

30) $x^3 \times 2y^3 x^4 \times 2y =$

31) $3yx^4 \times 3y^4 x \times 3xy^3 =$

32) $6y^3 \times 2y^2 x^4 \times 10yx^5 =$

WWW.MathNotion.Com

Zero and Negative Exponents

✎ Evaluate the following expressions.

1) $1^{-5} =$

2) $4^{-1} =$

3) $0^{10} =$

4) $1^{15} =$

5) $5^{-2} =$

6) $3^{-3} =$

7) $9^{-1} =$

8) $10^{-2} =$

9) $12^{-2} =$

10) $2^{-5} =$

11) $3^{-4} =$

12) $2^{-4} =$

13) $6^{-3} =$

14) $10^{-3} =$

15) $30^{-1} =$

16) $15^{-2} =$

17) $4^{-3} =$

18) $2^{-7} =$

19) $5^{-3} =$

20) $4^{-4} =$

21) $3^{-5} =$

22) $10^{-4} =$

23) $2^{-10} =$

24) $8^{-3} =$

25) $20^{-2} =$

26) $14^{-2} =$

27) $9^{-3} =$

28) $100^{-2} =$

29) $5^{-4} =$

30) $4^{-6} =$

31) $\left(\frac{1}{4}\right)^{-3} =$

32) $\left(\frac{1}{6}\right)^{-2} =$

33) $\left(\frac{1}{7}\right)^{-2} =$

34) $\left(\frac{2}{3}\right)^{-3} =$

35) $\left(\frac{1}{13}\right)^{-2} =$

36) $\left(\frac{7}{12}\right)^{-2} =$

37) $\left(\frac{1}{6}\right)^{-3} =$

38) $\left(\frac{1}{300}\right)^{-2} =$

39) $\left(\frac{2}{9}\right)^{-2} =$

40) $\left(\frac{7}{5}\right)^{-1} =$

41) $\left(\frac{13}{23}\right)^{0} =$

42) $\left(\frac{1}{4}\right)^{-5} =$

Division Property of Exponents

✎ Simplify.

1) $\dfrac{5^6}{5^7} =$

2) $\dfrac{8^8}{8^6} =$

3) $\dfrac{4^5}{4} =$

4) $\dfrac{3}{3^5} =$

5) $\dfrac{x}{x^6} =$

6) $\dfrac{3 \times 3^2}{3^2 \times 3^5} =$

7) $\dfrac{9^4}{9^2} =$

8) $\dfrac{10 \times 10^9}{10^2 \times 10^7} =$

9) $\dfrac{7^5 \times 7^7}{7^4 \times 7^8} =$

10) $\dfrac{15x}{30x^6} =$

11) $\dfrac{3x^9}{4x^4} =$

12) $\dfrac{15x^8}{10x^9} =$

13) $\dfrac{42x^5}{6y^9} =$

14) $\dfrac{36y^8}{4x^4 y^5} =$

15) $\dfrac{2x^7}{9x} =$

16) $\dfrac{49x^8 y^6}{7x^9} =$

17) $\dfrac{48x^2}{24x^6 y^{12}} =$

18) $\dfrac{30yx^5}{6yx^7} =$

19) $\dfrac{19x^7 y}{38x^{12} y^4} =$

20) $\dfrac{9x^8}{63x^8} =$

21) $\dfrac{9x^{-9}}{4x^{-3}} =$

Powers of Products and Quotients

✏️ Simplify.

1) $(4^3)^2 =$

2) $(2^3)^4 =$

3) $(2 \times 2^3)^2 =$

4) $(5 \times 5^5)^6 =$

5) $(19^4 \times 19^2)^3 =$

6) $(2^3 \times 2^4)^4 =$

7) $(5 \times 5^2)^2 =$

8) $(4^4)^4 =$

9) $(8x^5)^2 =$

10) $(3x^2 y^4)^4 =$

11) $(7x^5 y^2)^2 =$

12) $(5x^4 y^4)^3 =$

13) $(2x^3 y^3)^5 =$

14) $(10x^3 y^4)^3 =$

15) $(13y^3 y)^2 =$

16) $(5x^6 x^4)^2 =$

17) $(6x^7 y^6)^3 =$

18) $(12x^5 x^7)^2 =$

19) $(2x^4 \times 2x)^4 =$

20) $(2x^4 y^3)^5 =$

21) $(15x^7 y^2)^2 =$

22) $(8x^3 y^5)^3 =$

23) $(3x \times 2y^2)^4 =$

24) $\left(\dfrac{4x}{x^5}\right)^2 =$

25) $\left(\dfrac{x^4 y^5}{x^3 y^5}\right)^9 =$

26) $\left(\dfrac{36xy}{6x^5}\right)^3 =$

27) $\left(\dfrac{x^7}{x^8 y^2}\right)^6 =$

28) $\left(\dfrac{xy^4}{x^3 y^6}\right)^{-3} =$

29) $\left(\dfrac{5xy^8}{x^3}\right)^2 =$

30) $\left(\dfrac{xy^6}{2xy^3}\right)^{-4} =$

FSA Subject Test Mathematics Grade 7

Negative Exponents and Negative Bases

✎ **Simplify.**

1) $-9^{-1} =$

2) $-9^{-2} =$

3) $-2^{-5} =$

4) $-x^{-7} =$

5) $11x^{-1} =$

6) $-8x^{-3} =$

7) $-12x^{-5} =$

8) $-9x^{-8}y^{-6} =$

9) $32x^{-5}y^{-1} =$

10) $10a^{-9}b^{-3} =$

11) $-17x^4y^{-6} =$

12) $-\dfrac{25}{x^{-5}} =$

13) $-\dfrac{13x}{a^{-7}} =$

14) $\left(-\dfrac{1}{3}\right)^{-4} =$

15) $\left(-\dfrac{3}{4}\right)^{-2} =$

16) $-\dfrac{14}{a^{-6}b^{-3}} =$

17) $-\dfrac{7x}{x^{-8}} =$

18) $-\dfrac{a^{-9}}{b^{-5}} =$

19) $-\dfrac{11}{x^{-5}} =$

20) $\dfrac{8b}{-16c^{-6}} =$

21) $\dfrac{12ab}{a^{-4}b^{-3}} =$

22) $-\dfrac{8n^{-4}}{32p^{-7}} =$

23) $\dfrac{16ab^{-6}}{-6c^{-5}} =$

24) $\left(\dfrac{10a}{5c}\right)^{-4} =$

25) $\left(-\dfrac{12x}{4yz}\right)^{-3} =$

26) $\dfrac{8ab^{-7}}{-5c^{-3}} =$

27) $\left(-\dfrac{x^4}{x^5}\right)^{-5} =$

28) $\left(-\dfrac{x^{-2}}{7x^3}\right)^{-2} =$

29) $\left(-\dfrac{x^{-4}}{x^2}\right)^{-6} =$

Scientific Notation

✎ Write each number in scientific notation.

1) 0.223 =

2) 0.09 =

3) 4.5 =

4) 900 =

5) 2,000 =

6) 0.006 =

7) 33 =

8) 9,400 =

9) 1,470 =

10) 52,000 =

11) 8,000,000 =

12) 0.00009 =

13) 2,158,000 =

14) 0.0039 =

15) 0.000075 =

16) 4,300,000 =

17) 130,000 =

18) 4,000,000,000 =

19) 0.00009 =

20) 0.0039 =

✎ Write each number in standard notation.

21) 4×10^{-1} =

22) 1.2×10^{-3} =

23) 2.7×10^{5} =

24) 6×10^{-4} =

25) 3.6×10^{-3} =

26) 5.5×10^{5} =

27) 3.2×10^{4} =

28) 3.88×10^{6} =

29) 7×10^{-6} =

30) 4.2×10^{-7} =

FSA Subject Test Mathematics Grade 7

Square Roots

✎ Find the value each square root.

1) $\sqrt{16} = $ ___

2) $\sqrt{25} = $ ___

3) $\sqrt{1} = $ ___

4) $\sqrt{64} = $ ___

5) $\sqrt{0} = $ ___

6) $\sqrt{196} = $ ___

7) $\sqrt{4} = $ ___

8) $\sqrt{256} = $ ___

9) $\sqrt{36} = $ ___

10) $\sqrt{289} = $ ___

11) $\sqrt{169} = $ ___

12) $\sqrt{144} = $ ___

13) $\sqrt{100} = $ ___

14) $\sqrt{1,600} = $ ___

15) $\sqrt{2,500} = $ ___

16) $\sqrt{324} = $ ___

17) $\sqrt{529} = $ ___

18) $\sqrt{20} = $ ___

19) $\sqrt{625} = $ ___

20) $\sqrt{18} = $ ___

21) $\sqrt{50} = $ ___

22) $\sqrt{1,024} = $ ___

23) $\sqrt{160} = $ ___

24) $\sqrt{32} = $ ___

✎ Evaluate.

25) $\sqrt{4} \times \sqrt{25} = $ _____

26) $\sqrt{36} \times \sqrt{49} = $ _____

27) $\sqrt{6} \times \sqrt{6} = $ _____

28) $\sqrt{13} \times \sqrt{13} = $ _____

29) $2\sqrt{5} \times 3\sqrt{5} = $ _____

30) $\sqrt{12} \times \sqrt{3} = $ _____

31) $\sqrt{13} + \sqrt{13} = $ _____

32) $\sqrt{10} + 2\sqrt{10} = $ _____

33) $12\sqrt{7} - 10\sqrt{7} = $ _____

34) $4\sqrt{10} \times 2\sqrt{10} = $ _____

35) $5\sqrt{3} \times 8\sqrt{3} = $ _____

36) $6\sqrt{3} - \sqrt{12} = $ _____

WWW.MathNotion.Com

FSA Subject Test Mathematics Grade 7

Answers of Worksheets

Multiplication Property of Exponents

1) 4^6
2) 8^5
3) 7^6
4) 9^4
5) 2^7
6) 5^7
7) 4^7
8) $5x^2$
9) x^6
10) x^9
11) x^9
12) $30x^2$
13) $16x^6$
14) $7x^4$
15) $12x^6$
16) $5x^8$
17) $4x^9$
18) $6x^6$
19) $24x^8$
20) $20x^4y$
21) $7x^{10}y^5$
22) x^7y^7
23) $12x^8y^4$
24) $36x^6y^5$
25) $30x^{11}y^6$
26) $32x^4y^9$
27) $12x^4y^8$
28) $16x^7y^{10}$
29) $21x^3y^{13}$
30) $4x^7y^4$
31) $27x^6y^8$
32) $120x^9y^6$

Zero and Negative Exponents

1) 1
2) $\frac{1}{4}$
3) 0
4) 1
5) $\frac{1}{25}$
6) $\frac{1}{27}$
7) $\frac{1}{9}$
8) $\frac{1}{100}$
9) $\frac{1}{144}$
10) $\frac{1}{32}$
11) $\frac{1}{81}$
12) $\frac{1}{16}$
13) $\frac{1}{216}$
14) $\frac{1}{1,000}$
15) $\frac{1}{30}$
16) $\frac{1}{225}$
17) $\frac{1}{64}$
18) $\frac{1}{128}$
19) $\frac{1}{125}$
20) $\frac{1}{256}$
21) $\frac{1}{243}$
22) $\frac{1}{10,000}$
23) $\frac{1}{1,024}$
24) $\frac{1}{512}$
25) $\frac{1}{400}$
26) $\frac{1}{196}$
27) $\frac{1}{729}$
28) $\frac{1}{10,000}$
29) $\frac{1}{625}$
30) $\frac{1}{4,096}$
31) 64
32) 36
33) 49
34) $\frac{27}{8}$
35) 169
36) $\frac{144}{49}$
37) 216
38) $90,000$
39) $\frac{81}{4}$
40) $\frac{5}{7}$
41) 1
42) $1,024$

Division Property of Exponents

1) $\frac{1}{5}$
2) 8^2
3) 4^4
4) $\frac{1}{3^4}$
5) $\frac{1}{x^5}$
6) $\frac{1}{3^4}$
7) 9^2
8) 10
9) 1
10) $\frac{1}{2x^5}$
11) $\frac{3x^5}{4}$
12) $\frac{3}{2x}$
13) $\frac{7x^5}{y^9}$
14) $\frac{9y^3}{x^4}$
15) $\frac{2x^6}{9}$

WWW.MathNotion.Com

FSA Subject Test Mathematics Grade 7

16) $\dfrac{7y^6}{x}$

17) $\dfrac{2}{x^4 y^{12}}$

18) $\dfrac{5}{x^2}$

19) $\dfrac{1}{2x^5 y^3}$

20) $\dfrac{1}{7}$

21) $\dfrac{9}{4x^6}$

Powers of Products and Quotients

1) 4^6
2) 2^{12}
3) 2^8
4) 5^{36}
5) 19^{18}
6) 2^{28}
7) 5^6
8) 4^{16}
9) $64x^{10}$
10) $81x^8 y^{16}$
11) $49x^{10} y^4$
12) $125x^{12} y^{12}$
13) $32x^{15} y^{15}$
14) $1{,}000 x^9 y^{12}$
15) $169 y^8$
16) $25 x^{20}$
17) $216 x^{21} y^{18}$
18) $144 x^{24}$
19) $256 x^{20}$
20) $32 x^{20} y^{15}$
21) $225 x^{14} y^4$
22) $512 x^9 y^{15}$
23) $1{,}296 x^4 y^8$
24) $\dfrac{16}{x^8}$
25) x^9
26) $\dfrac{216 y^3}{x^{12}}$
27) $\dfrac{1}{x^6 y^{12}}$
28) $x^6 y^6$
29) $\dfrac{25 y^{16}}{x^4}$
30) $\dfrac{16}{y^{12}}$

Negative Exponents and Negative Bases

1) $-\dfrac{1}{9}$
2) $-\dfrac{1}{81}$
3) $-\dfrac{1}{32}$
4) $-\dfrac{1}{x^7}$
5) $\dfrac{11}{x}$
6) $-\dfrac{8}{x^3}$
7) $-\dfrac{12}{x^5}$
8) $-\dfrac{9}{x^8 y^6}$
9) $\dfrac{32}{x^5 y}$
10) $\dfrac{10}{a^9 b^3}$
11) $-\dfrac{17 x^4}{y^6}$
12) $-25 x^5$
13) $-13 x a^7$
14) 81
15) $\dfrac{16}{9}$
16) $-14 a^6 b^3$
17) $-7 x^9$
18) $-\dfrac{b^5}{a^9}$
19) $-11 x^5$
20) $-\dfrac{bc^6}{2}$
21) $12 a^5 b^4$
22) $-\dfrac{p^7}{4 n^4}$
23) $-\dfrac{8 a c^5}{3 b^6}$
24) $\dfrac{c^4}{16 a^4}$
25) $\dfrac{y^3 z^3}{27 x^3}$
26) $-\dfrac{8 a c^3}{5 b^7}$
27) $-x^5$
28) $49 x^{10}$
29) x^{36}

Scientific Notation

FSA Subject Test Mathematics Grade 7

1) 2.23×10^{-1}
2) 9×10^{-2}
3) 4.5×10^0
4) 9×10^2
5) 2×10^3
6) 6×10^{-3}
7) 3.3×10^1
8) 9.4×10^3
9) 1.47×10^3
10) 5.2×10^4
11) 8×10^6
12) 9×10^{-5}
13) 2.158×10^6
14) 3.9×10^{-3}
15) 7.5×10^{-5}
16) 4.3×10^6
17) 1.3×10^5
18) 4×10^9
19) 9×10^{-5}
20) 3.9×10^{-3}
21) 0.4
22) 0.0012
23) 270,000
24) 0.0006
25) 0.0036
26) 550,000
27) 32,000
28) 3,880,000
29) 0.000007
30) 0.00000042

Square Roots

1) 4
2) 5
3) 1
4) 8
5) 0
6) 14
7) 2
8) 16
9) 6
10) 17
11) 13
12) 12
13) 10
14) 40
15) 50
16) 18
17) 23
18) $2\sqrt{5}$
19) 25
20) $3\sqrt{2}$
21) $5\sqrt{2}$
22) 32
23) $4\sqrt{10}$
24) $4\sqrt{2}$
25) 10
26) 42
27) 6
28) 13
29) 30
30) 6
31) $2\sqrt{13}$
32) $3\sqrt{10}$
33) $2\sqrt{7}$
34) 80
35) 120
36) $4\sqrt{3}$

WWW.MathNotion.Com

Chapter 6:
Algebraic Expressions

Topics that you will practice in this chapter:

- ✓ Translate Phrases into an Algebraic Statement
- ✓ Simplifying Variable Expressions
- ✓ The Distributive Property
- ✓ Evaluating One Variable Expressions
- ✓ Evaluating Two Variables Expressions
- ✓ Combining like Terms

Mathematics is, as it were, a sensuous logic, and relates to philosophy as do the arts, music, and plastic art to poetry. — K. Shegel

FSA Subject Test Mathematics Grade 7

Translate Phrases into an Algebraic Statement

✏ Write an algebraic expression for each phrase.

1) 9 multiplied by x. _____

2) Subtract 11 from y. _____

3) 19 divided by x. _____

4) 38 decreased by y. _____

5) Add y to 40. _____

6) The square of 6. _____

7) x raised to the fifth power. _____

8) The sum of six and a number. _____

9) The difference between fifty-seven and y. _____

10) The quotient of nine and a number. _____

11) The quotient of the square of x and 25. _____

12) The difference between x and 6 is 19. _____

13) 10 times a reduced by the square of b. _____

14) Subtract the product of a and b from 41. _____

WWW.MathNotion.Com

FSA Subject Test Mathematics Grade 7

Simplifying Variable Expressions

✏️ **Simplify each expression.**

1) $3(x + 5) =$

2) $(-4)(7x - 5) =$

3) $11x + 5 - 6x =$

4) $-4 - 2x^2 - 6x^2 =$

5) $7 + 13x^2 + 3 =$

6) $3x^2 + 7x + 15x^2 =$

7) $3x^2 - 12x^2 + 4x =$

8) $4x^2 - 8x - 2x =$

9) $6x + 7(3 - 4x) =$

10) $8x + 4(15x - 3) =$

11) $6(-3x - 9) - 17 =$

12) $-11x^2 - (-5x) =$

13) $2x + 7 + 5 - 8x =$

14) $7 + 6x - 11 - 5x =$

15) $27x + 8 - 13 - 5x =$

16) $(-11)(-5x + 2) - 41x =$

17) $19x - 4(4 - 2x) =$

18) $16x + 3(3x + 6) + 10 =$

19) $5(-2x - 4) - 13x =$

20) $16x - 3x(x + 10) =$

21) $17x + 5x(2 - 4x) =$

22) $5x(-4x - 7) + 20x =$

23) $25x - 19 + 4x^2 =$

24) $6x(x - 11) + 25 =$

25) $4x - 5 + 15x + 3x^2 =$

26) $-7x^2 - 11x - 9x =$

27) $10x - 9x^2 - 3x^2 - 7 =$

28) $13 + 3x^2 - 9x^2 - 21x =$

29) $22x + 10x^2 - 15x + 17 =$

30) $4x^2 + 25x + 21x^2 =$

31) $29 - 12x^2 - 23x - 4x^2 =$

32) $22x - 19x - 9x^2 + 30 =$

WWW.MathNotion.Com

FSA Subject Test Mathematics Grade 7

The Distributive Property

✎ Use the distributive property to simply each expression.

1) $4(1 + 2x) =$

2) $2(4 + 7x) =$

3) $3(4x − 4) =$

4) $(2x − 5)(−6) =$

5) $(−3)(x + 6) =$

6) $(4 + 3x)2 =$

7) $(−5)(8 − 3x) =$

8) $−(−5 − 7x) =$

9) $(−6x + 3)(−3) =$

10) $(−4)(x − 7) =$

11) $−(5 − 3x) =$

12) $3(9 + 4x) =$

13) $6(4 + 3x) =$

14) $(−5x + 3)2 =$

15) $(5 − 8x)(−3) =$

16) $(−12)(3x + 3) =$

17) $(5 − 3x)6 =$

18) $4(2 + 6x) =$

19) $8(7x − 3) =$

20) $(−2x + 3)4 =$

21) $(7 − 5x)(−9) =$

22) $(−10)(x − 8) =$

23) $(11 − 4x)3 =$

24) $(−6)(10x − 4) =$

25) $(3 − 9x)(−7) =$

26) $(−9)(x + 9) =$

27) $(−3 + 5x)(−7) =$

28) $(−5)(8 − 10x) =$

29) $12(4x − 8) =$

30) $(−10x + 13)(−3) =$

31) $(−8)(3x − 2) + 4(x + 5) =$

32) $(−8)(x + 4) − (6 + 5x) =$

WWW.MathNotion.Com

FSA Subject Test Mathematics Grade 7

Evaluating One Variable Expressions

✎ **Evaluate each expression using the value given.**

1) $8 - x$, $x = 5$

2) $x - 9$, $x = 5$

3) $5x + 4$, $x = 3$

4) $x - 13$, $x = -4$

5) $12 - x$, $x = 4$

6) $x + 2$, $x = 6$

7) $4x + 8$, $x = 3$

8) $x + (-7)$, $x = -8$

9) $4x + 5$, $x = 2$

10) $3x + 9$, $x = -2$

11) $15 + 3x - 7$, $x = 2$

12) $17 - 3x$, $x = 3$

13) $8x - 9$, $x = 4$

14) $5x + 4$, $x = -3$

15) $10x + 5$, $x = 3$

16) $14 - 4x$, $x = -6$

17) $3(5x + 3)$, $x = 9$

18) $4(-3x - 6)$, $x = 3$

19) $7x - 2x + 12$, $x = 4$

20) $(5x + 6) \div 2$, $x = 8$

21) $(x + 18) \div 10$, $x = 12$

22) $5x - 12 + 3x$, $x = -3$

23) $(6 - 4x)(-3)$, $x = -4$

24) $9x^2 + 3x - 6$, $x = 2$

25) $x^2 - 10x$, $x = -5$

26) $3x(7 - 2x)$, $x = 2$

27) $12x + 6 - 2x^2$, $x = -4$

28) $(-3)(4x - 8 + 3x)$, $x = 3$

29) $(-6) + \frac{x}{4} + 3x$, $x = 16$

30) $(-6) + \frac{x}{5}$, $x = 35$

31) $\left(-\frac{45}{x}\right) - 7 + 2x$, $x = 9$

32) $\left(-\frac{21}{x}\right) - 12 + 4x$, $x = 7$

WWW.MathNotion.Com

Evaluating Two Variables Expressions

✍ Evaluate each expression using the values given.

1) $2x - 4y$,
   $x = 4, y = 1$

2) $3x + 5y$,
   $x = -2, y = 2$

3) $-7a + 4b$,
   $a = 2, b = 4$

4) $3x + 5 - y$,
   $x = 5, y = 6$

5) $3z + 12 - 2k$,
   $z = 5, k = 6$

6) $6(-x - 3y)$,
   $x = 5, y = -2$

7) $5a + 3b$,
   $a = 3, b = 4$

8) $7x \div 3y$,
   $x = 3, y = 7$

9) $2x + 15 + 5y$,
   $x = -3, y = 1$

10) $5a - (18 - b)$,
    $a = 2, b = 8$

11) $2z + 20 + 5k$,
    $z = -6, k = 5$

12) $xy + 10 + 4x$,
    $x = 3, y = 5$

13) $2x + 4y - 8 + 5$,
    $x = 5, y = 2$

14) $\left(-\frac{24}{x}\right) + 3 + 2y$,
    $x = 4, y = 6$

15) $(-3)(-3a - 3b)$,
    $a = 4, b = 5$

16) $12 + 4x - 7 - y$,
    $x = 3, y = 5$

17) $11x + 5 - 8y + 6$,
    $x = 5, y = 2$

18) $10 + 2(-4x - 5y)$,
    $x = 5, y = 4$

19) $5x + 13 + 6y$,
    $x = 5, y = 6$

20) $10a - (7a + 3b) - 11$,
    $a = 3, b = 8$

FSA Subject Test Mathematics Grade 7

Combining like Terms

✏️ **Simplify each expression.**

1) $11x + 3x + 6 =$

2) $8(2x - 6) =$

3) $18x - 7x + 11 =$

4) $(-4)(6x - 7) =$

5) $22x - 10x - 5 =$

6) $32x - 13 + 8x =$

7) $15 - (8x - 11) =$

8) $-24x + 17 - 11x =$

9) $12x - 8 - 6x + 9 =$

10) $21x + 5 - 36 + 12x =$

11) $28x + 3x - 11 =$

12) $(-3x + 4)5 =$

13) $2 + 4x + 9x - 8 =$

14) $6(2x - 5x) - 4 =$

15) $4(5x + 11) + 3x =$

16) $x - 14 - 11x =$

17) $5(10 + 9x) - 8x =$

18) $42x + 17 - 23x =$

19) $(-7x) + 19 + 20x =$

20) $(-7x) - 33 + 29x =$

21) $4(5x + 3) - 19x =$

22) $5(6 - 2x) - 15x =$

23) $-24x + (11 - 18x) =$

24) $(-9) - (6)(7x + 3) =$

25) $(-1)(8x - 10) - 21x =$

26) $-36x + 14 + 27x - 5x =$

27) $3(-13x + 6) - 17x =$

28) $-5x - 42 + 32x =$

29) $37x - 19x + 15 - 9x =$

30) $3(5x + 7x) - 31 =$

31) $14 - 6x - 15 - 9x =$

32) $-2(-5x - 7x) + 27x =$

WWW.MathNotion.Com

FSA Subject Test Mathematics Grade 7

Answers of Worksheets

Translate Phrases into an Algebraic Statement

1) $9x$
2) $y - 11$
3) $\frac{19}{x}$
4) $38 - y$
5) $y + 40$
6) 6^2
7) x^5
8) $6 + x$
9) $57 - y$
10) $\frac{9}{x}$
11) $\frac{x^2}{25}$
12) $x - 6 = 19$
13) $10a - b^2$
14) $41 - ab$

Simplifying Variable Expressions

1) $3x + 15$
2) $-28x + 20$
3) $5x + 5$
4) $-8x^2 - 4$
5) $13x^2 + 10$
6) $18x^2 + 7x$
7) $-9x^2 + 4x$
8) $4x^2 - 10x$
9) $-22x + 21$
10) $68x - 12$
11) $-18x - 71$
12) $-11x^2 + 5x$
13) $-6x + 12$
14) $x - 4$
15) $22x - 5$
16) $14x - 22$
17) $27x - 16$
18) $25x + 28$
19) $-23x - 20$
20) $-3x^2 - 14x$
21) $-20x^2 + 27x$
22) $-20x^2 - 15x$
23) $4x^2 + 25x - 19$
24) $6x^2 - 66x + 25$
25) $3x^2 + 19x - 5$
26) $-7x^2 - 20x$
27) $-12x^2 + 10x - 7$
28) $-6x^2 - 21x + 13$
29) $10x^2 + 7x + 17$
30) $25x^2 + 25x$
31) $-16x^2 - 23x + 29$
32) $-9x^2 + 3x + 30$

The Distributive Property

1) $8x + 4$
2) $14x + 8$
3) $12x - 12$
4) $-12x + 30$
5) $-3x - 18$
6) $6x + 8$
7) $15x - 40$
8) $7x + 5$
9) $18x - 9$
10) $-4x + 28$
11) $3x - 5$
12) $12x + 27$
13) $18x + 24$
14) $-10x + 6$
15) $24x - 15$
16) $-36x - 36$
17) $-18x + 30$
18) $24x + 8$
19) $56x - 24$
20) $-8x + 12$
21) $45x - 63$
22) $-10x + 80$
23) $-12x + 33$
24) $-60x + 24$
25) $63x - 21$
26) $-9x - 81$
27) $-35x + 21$
28) $50x - 40$
29) $48x - 96$
30) $30x - 39$
31) $-20x + 36$
32) $-13x - 38$

Evaluating One Variables

1) 3
2) −4
3) 19
4) −17
5) 8
6) 8
7) 20
8) −15
9) 13
10) 3
11) 14
12) 8

WWW.MathNotion.Com

FSA Subject Test Mathematics Grade 7

13) 23
14) −11
15) 35
16) 38
17) 144

18) −60
19) 32
20) 23
21) 3
22) −36

23) −66
24) 36
25) 75
26) 18
27) −74

28) −39
29) 46
30) 1
31) 6
32) 13

Evaluating Two Variables

1) 4
2) 4
3) 2
4) 14
5) 15

6) 6
7) 27
8) 1
9) 14
10) 0

11) 33
12) 37
13) 15
14) 9
15) 81

16) 12
17) 50
18) −70
19) 74
20) −26

Combining like Terms

1) $14x + 6$
2) $16x - 48$
3) $11x + 11$
4) $-24x + 28$
5) $12x - 5$
6) $40x - 13$
7) $-8x + 26$
8) $-35x + 17$

9) $6x + 1$
10) $33x - 31$
11) $31x - 11$
12) $-15x + 20$
13) $13x - 6$
14) $-18x - 4$
15) $23x + 44$
16) $-10x - 14$

17) $37x + 50$
18) $19x + 17$
19) $13x + 19$
20) $22x - 33$
21) $x + 12$
22) $-25x + 30$
23) $-42x + 11$
24) $-42x - 27$

25) $-29x + 10$
26) $-14x + 14$
27) $-56x + 18$
28) $27x - 42$
29) $9x + 15$
30) $36x - 31$
31) $-15x - 1$
32) $51x$

WWW.MathNotion.Com

Chapter 7:
Equations and Inequalities

Topics that you will practice in this chapter:

- ✓ One–Step Equations
- ✓ Multi–Step Equations
- ✓ Graphing Single–Variable Inequalities
- ✓ One–Step Inequalities
- ✓ Multi-Step Inequalities
- ✓ Systems of Equations
- ✓ Systems of Equations Word Problems

"Life is a math equation. In order to gain the most, you have to know how to convert negatives into positives." – Anonymous

FSA Subject Test Mathematics Grade 7

One–Step Equations

✏ **Find the answer for each equation.**

1) $3x = 90$, $x =$ ____

2) $5x = 35$, $x =$ ____

3) $6x = 24$, $x =$ ____

4) $24x = 144$, $x =$ ____

5) $x + 15 = 20$, $x =$ ____

6) $x - 7 = 4$, $x =$ ____

7) $x - 9 = 2$, $x =$ ____

8) $x + 15 = 23$, $x =$ ____

9) $x - 4 = 13$, $x =$ ____

10) $12 = 16 + x$, $x =$ ____

11) $x - 10 = 2$, $x =$ ____

12) $5 - x = -11$, $x =$ ____

13) $28 = -6 + x$, $x =$ ____

14) $x - 20 = -35$, $x =$ ____

15) $x + 14 = -4$, $x =$ ____

16) $14 = 28 - x$, $x =$ ____

17) $7 + x = -7$, $x =$ ____

18) $x - 16 = 4$, $x =$ ____

19) $30 = x - 15$, $x =$ ____

20) $x - 5 = -18$, $x =$ ____

21) $x - 10 = 24$, $x =$ ____

22) $x - 20 = -25$, $x =$ ____

23) $x - 17 = 30$, $x =$ ____

24) $-70 = x - 28$, $x =$ ____

25) $x - 9 = 13$, $x =$ ____

26) $36 = 4x$, $x =$ ____

27) $x - 35 = 25$, $x =$ ____

28) $x - 25 = 10$, $x =$ ____

29) $70 - x = 16$, $x =$ ____

30) $x - 10 = 14$, $x =$ ____

31) $17 - x = -13$, $x =$ ____

32) $x - 9 = -30$, $x =$ ____

WWW.MathNotion.Com

FSA Subject Test Mathematics Grade 7

Multi-Step Equations

✎ Find the answer for each equation.

1) $3x + 3 = 9$

2) $-x + 5 = 12$

3) $4x - 8 = 8$

4) $-(3 - x) = 5$

5) $4x - 8 = 16$

6) $12x - 15 = 9$

7) $2x - 18 = 2$

8) $4x + 8 = 16$

9) $24x + 27 = 75$

10) $-14(3 + x) = 14$

11) $-3(2 + x) = 6$

12) $12 = -(x - 7)$

13) $3(3 - x) = 30$

14) $-15 = -(3x + 6)$

15) $40(3 + x) = 40$

16) $5(x - 10) = 25$

17) $-18 = x + 8x$

18) $3x + 25 = -2x - 10$

19) $7(6 + 3x) = -63$

20) $18 - 3x = -4 - 5x$

21) $4 - 6x = 36 + 2x$

22) $15 + 15x = -5 + 5x$

23) $42 = (-6x) - 7 + 7$

24) $21 = 3x - 21 + 4x$

25) $-18 = -6x - 9 + 3x$

26) $5x - 15 = -29 + 6x$

27) $7x - 18 = 4x + 3$

28) $-7 - 4x = 5(4 - x)$

29) $x - 5 = -5(-3 - x)$

30) $13x - 68 = 15x - 102$

31) $-5x - 3 = -3(9 + 3x)$

32) $-2x - 15 = 6x + 17$

WWW.MathNotion.Com

FSA Subject Test Mathematics Grade 7

Graphing Single–Variable Inequalities

✎ **Draw a graph for each inequality.**

1) $x > -1$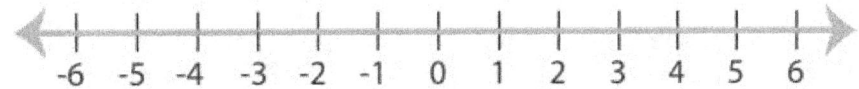

2) $x \leq 2$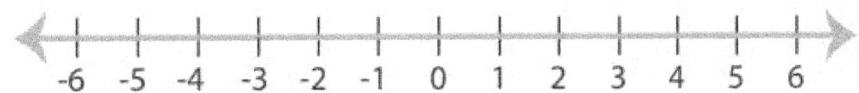

3) $x \geq 0$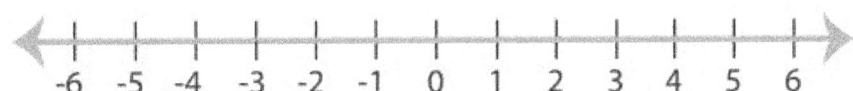

4) $x < -3$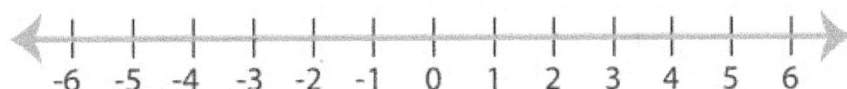

5) $x < \frac{1}{2}$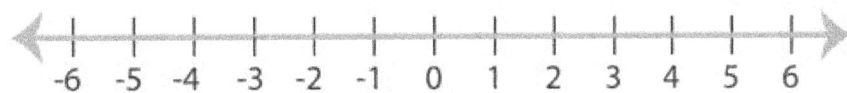

6) $x \leq -2$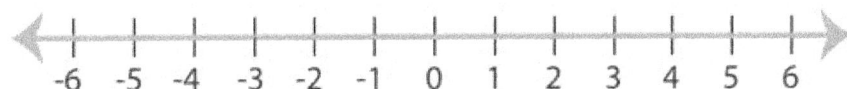

7) $x \leq 3$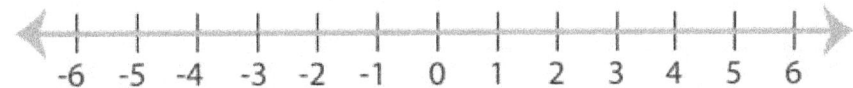

8) $x \geq -\frac{7}{2}$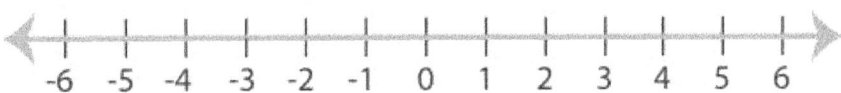

WWW.MathNotion.Com

FSA Subject Test Mathematics Grade 7

One–Step Inequalities

✏ **Find the answer for each inequality and graph it.**

1) $x + 4 \geq 4$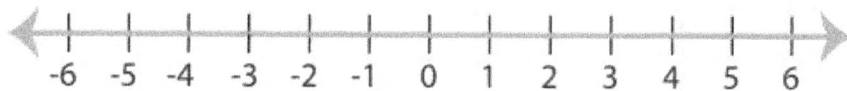

2) $x - 5 \leq 2$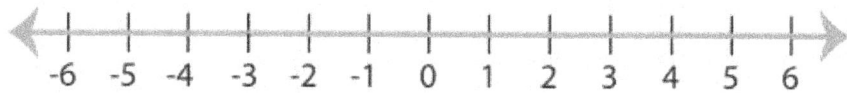

3) $5x > 35$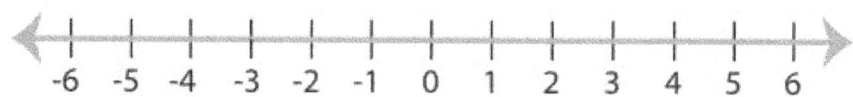

4) $9 + x \leq 11$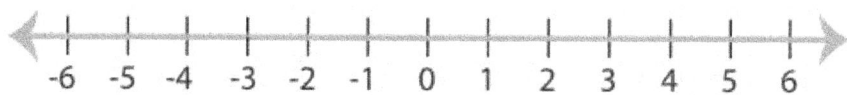

5) $x - 5 < -9$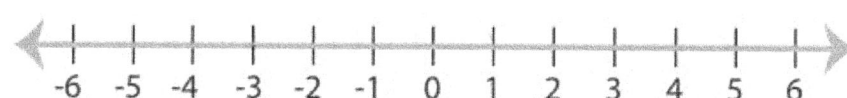

6) $9x \geq 72$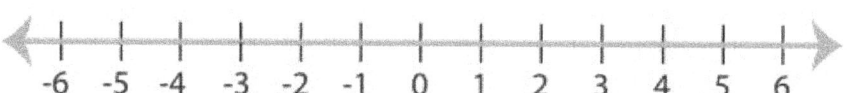

7) $9x \leq 27$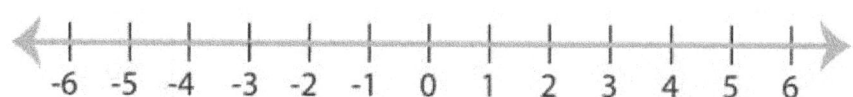

8) $x + 19 > 16$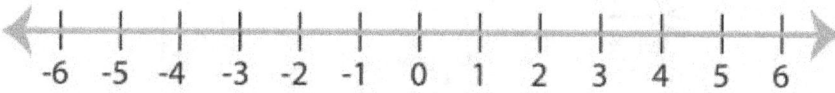

FSA Subject Test Mathematics Grade 7

Multi-Step Inequalities

✎ **Calculate each inequality.**

1) $x - 3 \leq 7$

2) $8 - x \leq 8$

3) $3x - 9 \leq 9$

4) $4x - 4 \geq 8$

5) $x - 7 \geq 1$

6) $5x - 15 \leq 5$

7) $6x - 8 \leq 4$

8) $-11 + 6x \leq 12$

9) $4(x - 4) \leq 16$

10) $3x - 10 \leq 11$

11) $5x - 25 < 25$

12) $9x - 5 < 22$

13) $20 - 7x \geq -15$

14) $33 + 6x < 45$

15) $8 + 8x \geq 96$

16) $7 + 3x < 13$

17) $4x - 3 < 9$

18) $5(2 - 2x) \geq -30$

19) $-(7 + 6x) < 29$

20) $12 - 8x \geq -20$

21) $-4(x - 6) > 24$

22) $\dfrac{3x + 9}{6} \leq 10$

23) $\dfrac{4x - 10}{3} \leq 2$

24) $\dfrac{2x - 8}{3} > 2$

25) $8 + \dfrac{x}{6} < 9$

26) $\dfrac{9x}{7} - 4 < 5$

27) $\dfrac{15x + 45}{15} > 1$

28) $16 + \dfrac{x}{4} < 6$

WWW.MathNotion.Com

FSA Subject Test Mathematics Grade 7

Systems of Equations

✏ **Calculate each system of equations.**

1) $-x + y = 2$ $x = $ ___
   $-4x + 2y = 6$ $y = $ ___

2) $-15x + 3y = -9$ $x = $ ___
   $9x - 16y = 48$ $y = $ ___

3) $y = -7$ $x = $ ___
   $6x + 5y = 7$ $y = $ ___

4) $3y = -9x + 15$ $x = $ ___
   $5x - 4y = -3$ $y = $ ___

5) $10x - 9y = -13$ $x = $ ___
   $-5x + 3y = 11$ $y = $ ___

6) $-12x - 16y = 20$ $x = $ ___
   $6x - 12y = 30$ $y = $ ___

7) $5x - 14y = -23$ $x = $ ___
   $-18x + 21y = 24$ $y = $ ___

8) $15x - 21y = -6$ $x = $ ___
   $2x - 3y = -2$ $y = $ ___

9) $-x + 3y = 3$ $x = $ ___
   $-14x + 16y = -10$ $y = $ ___

10) $x + 5y = 50$ $x = $ ___
    $3x + 10y = 80$ $y = $ ___

11) $6x - 7y = -8$ $x = $ ___
    $-x - 4y = -9$ $y = $ ___

12) $2x + 4y = -10$ $x = $ ___
    $2x - 8y = 14$ $y = $ ___

13) $4x + 3y = 12$ $x = $ ___
    $5x - 3y = 15$ $y = $ ___

14) $3x - 2y = 3$ $x = $ ___
    $7x - 8y = 22$ $y = $ ___

15) $3x + 2y = 5$ $x = $ ___
    $-10x - 4y = -14$ $y = $ ___

16) $10x + 7y = 1$ $x = $ ___
    $-5x - 7y = 24$ $y = $ ___

WWW.MathNotion.Com

FSA Subject Test Mathematics Grade 7

Systems of Equations Word Problems

✎ Find the answer for each word problem.

1) Tickets to a movie cost $4 for adults and $3 for students. A group of friends purchased 8 tickets for $31.00. How many adults ticket did they buy? ____

2) At a store, Eva bought two shirts and five hats for $77.00. Nicole bought three same shirts and four same hats for $84.00. What is the price of each shirt? ____

3) A farmhouse shelters 18 animals, some are pigs, and some are ducks. Altogether there are 66 legs. How many pigs are there? ____

4) A class of 214 students went on a field trip. They took 36 vehicles, some cars and some buses. If each car holds 5 students and each bus hold 22 students, how many buses did they take? ____

5) A theater is selling tickets for a performance. Mr. Smith purchased 5 senior tickets and 3 child tickets for $105 for his friends and family. Mr. Jackson purchased 3 senior tickets and 5 child tickets for $79. What is the price of a senior ticket? $____

6) The difference of two numbers is 10. Their sum is 20. What is the bigger number? $____

7) The sum of the digits of a certain two-digit number is 7. Reversing its digits increase the number by 9. What is the number? ____

8) The difference of two numbers is 11. Their sum is 25. What are the numbers? _____

9) The length of a rectangle is 5 meters greater than 2 times the width. The perimeter of rectangle is 28 meters. What is the length of the rectangle? _____

10) Jim has 25 nickels and dimes totaling $1.80. How many nickels does he have? ____

WWW.MathNotion.Com

FSA Subject Test Mathematics Grade 7

Answers of Worksheets

One–Step Equations

1) 30	9) 17	17) −14	25) 22
2) 7	10) −4	18) 20	26) 9
3) 4	11) 12	19) 45	27) 60
4) 6	12) 16	20) −13	28) 35
5) 5	13) 34	21) 34	29) 54
6) 11	14) −15	22) −5	30) 24
7) 11	15) −18	23) 47	31) 30
8) 8	16) 14	24) −42	32) −21

Multi–Step Equations

1) 2	9) 2	17) −2	25) 3
2) −7	10) −4	18) −7	26) 14
3) 4	11) −4	19) −5	27) 7
4) 8	12) −5	20) −11	28) 27
5) 6	13) −7	21) −4	29) −5
6) 2	14) 3	22) −2	30) 17
7) 10	15) −2	23) −7	31) −6
8) 2	16) 15	24) 6	32) −4

Graphing Single–Variable Inequalities

1) [number line with open circle at −1, shaded to the right]

2) [number line with closed circle at 2, shaded to the left]

3) [number line with closed circle at 0, shaded to the right]

4) [number line with open circle at −3, shaded to the right]

WWW.MathNotion.Com

FSA Subject Test Mathematics Grade 7

5)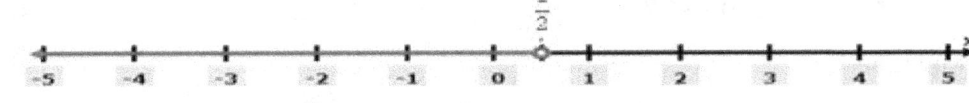

6)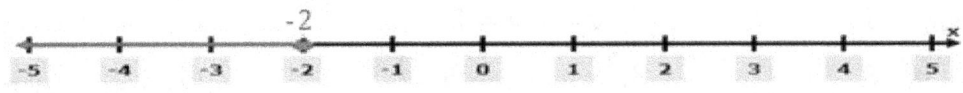

7)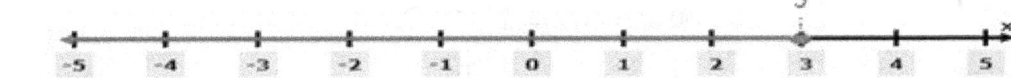

8)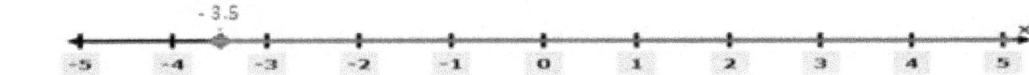

One–Step Inequalities

1)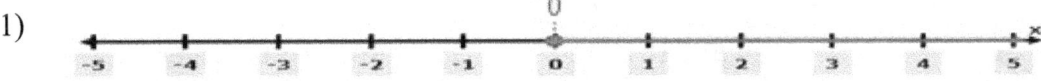

2)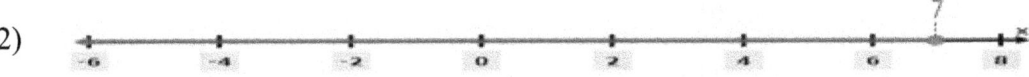

3)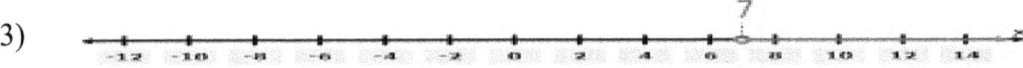

4)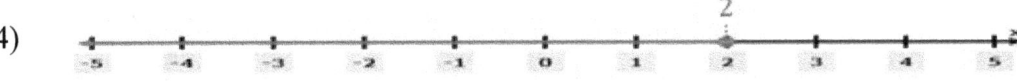

5)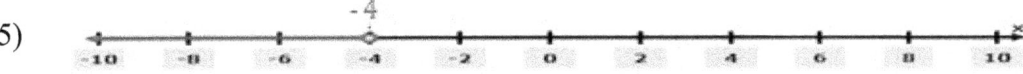

6)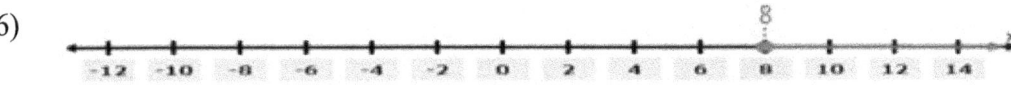

7)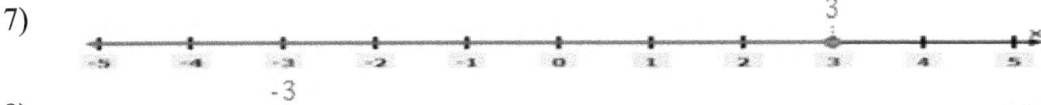

8)

Multi-Step Inequalities

1) $x \leq 10$
2) $x \geq 0$
3) $x \leq 6$
4) $x \geq 3$
5) $x \geq 8$
6) $x \leq 4$
7) $x \leq 2$
8) $x \leq \frac{23}{6}$
9) $x \leq 8$
10) $x \leq 7$
11) $x < 10$
12) $x < 3$
13) $x \leq 5$
14) $x < 2$
15) $x \geq 11$
16) $x < 2$
17) $x < 3$
18) $x \leq 4$
19) $x > -6$
20) $x \leq 4$
21) $x < 0$
22) $x \leq 17$
23) $x \leq 4$
24) $x > 7$

WWW.MathNotion.Com

FSA Subject Test Mathematics Grade 7

25) $x < 6$ 26) $x < 7$ 27) $x > -2$ 28) $x < -40$

Systems of Equations

1) $x = -1, y = 1$ 7) $x = 1, y = 2$ 13) $x = 3, y = 0$
2) $x = 0, y = -3$ 8) $x = 8, y = 6$ 14) $x = -2, y = -\frac{9}{2}$
3) $x = 7$ 9) $x = 3, y = 2$ 15) $x = 1, y = 1$
4) $x = 1, y = 2$ 10) $x = -20, y = 14$ 16) $x = 5, y = -7$
5) $x = -4, y = -3$ 11) $x = 1, y = 2$
6) $x = 1, y = -2$ 12) $x = -1, y = -2$

Systems of Equations Word Problems

1) 7   5) $18   9) 11 meters
2) $16   6) 15   10) 14
3) 15   7) 34
4) 2   8) 18, 7

FSA Subject Test Mathematics Grade 7

Chapter 8 :
Linear Functions

Topics that you will practice in this chapter:

- ✓ Finding Slope
- ✓ Graphing Lines Using Line Equation
- ✓ Writing Linear Equations
- ✓ Graphing Linear Inequalities
- ✓ Finding Midpoint
- ✓ Finding Distance of Two Points

"Nature is written in mathematical language." – Galileo Galilei

FSA Subject Test Mathematics Grade 7

Finding Slope

✏️ Find the slope of each line.

1) $y = x + 8$

2) $y = -3x + 5$

3) $y = 2x + 12$

4) $y = -4x + 19$

5) $y = 11 + 6x$

6) $y = 7 - 5x$

7) $y = 8x + 19$

8) $y = -9x + 20$

9) $y = -7x + 4$

10) $y = 3x - 8$

11) $y = \frac{1}{3}x + 8$

12) $y = -\frac{4}{5}x + 9$

13) $-3x + 6y = 30$

14) $4x + 4y = 16$

15) $3y - x = 10$

16) $8y - x = 5$

✏️ Find the slope of the line through each pair of points.

17) $(2, 3), (7, 10)$

18) $(-3, 5), (2, 15)$

19) $(5, -3), (1, 9)$

20) $(-5, -5), (10, 25)$

21) $(22, 3), (7, 18)$

22) $(-16, 8), (-7, 26)$

23) $(25, 11), (29, 19)$

24) $(26, -19), (14, 17)$

25) $(22, -13), (20, -11)$

26) $(19, 7), (15, -3)$

27) $(5, 7), (11, 19)$

28) $(52, -62), (40, 70)$

FSA Subject Test Mathematics Grade 7

Graphing Lines Using Line Equation

✏️ **Sketch the graph of each line.**

1) $y = x - 2$

2) $y = -3x + 2$

3) $x + y = 0$

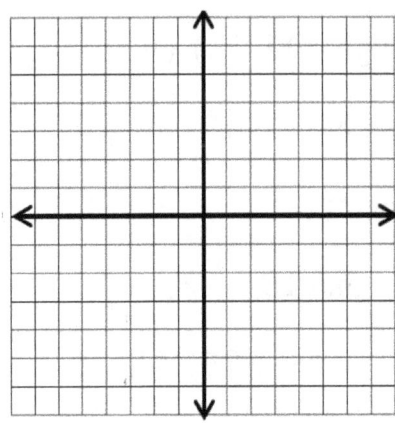 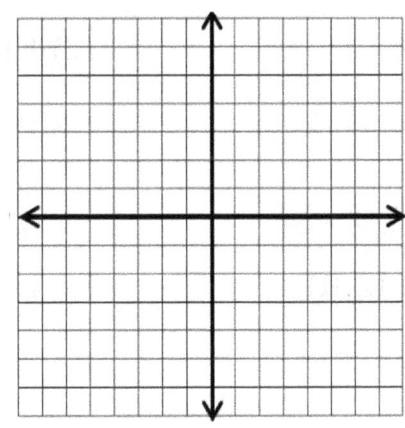 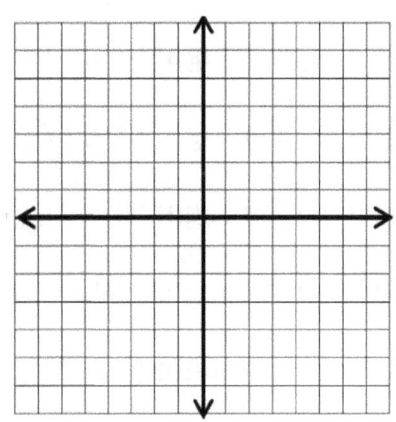

4) $x + y = -3$

5) $2x + 3y = -4$

6) $y - 3x + 6 = 0$

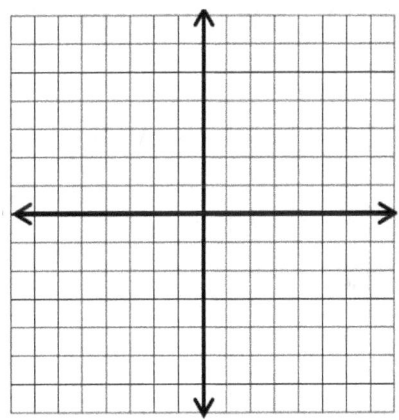 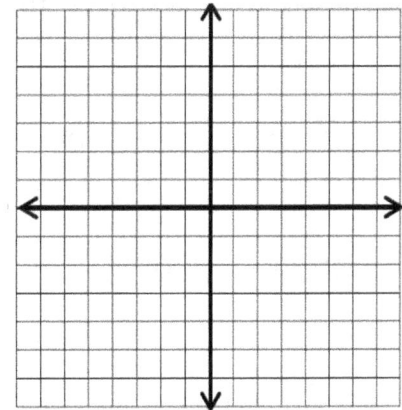 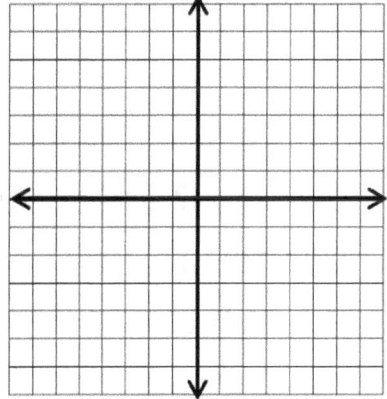

WWW.MathNotion.Com

FSA Subject Test Mathematics Grade 7

Writing Linear Equations

✍ **Write the equation of the line through the given points.**

1) Through: $(2, -5), (3, 9)$

2) Through: $(-6, 3), (3, 12)$

3) Through: $(10, 7), (5, 27)$

4) Through: $(15, 11), (3, -1)$

5) Through: $(24, 17), (12, -7)$

6) Through: $(8, 29), (4, -7)$

7) Through: $(20, -16), (12, 0)$

8) Through: $(-3, 10), (2, -5)$

9) Through: $(-6, 17), (4, -3)$

10) Through: $(-8, 22), (5, -4)$

11) Through: $(9, 27), (3, -3)$

12) Through: $(11, 32), (9, 4)$

13) Through: $(-3, 13), (-4, 0)$

14) Through: $(-5, 5), (5, 15)$

15) Through: $(18, -32), (11, 3)$

16) Through: $(-4, 25), (4, -15)$

✍ **Find the answer for each problem.**

17) What is the equation of a line with slope 6 and intercept 12? _____

18) What is the equation of a line with slope -11 and intercept -4? _____

19) What is the equation of a line with slope -3 and passes through point $(5, 2)$? _____

20) What is the equation of a line with slope -5 and passes through point $(-2, -1)$? _____

21) The slope of a line is -10 and it passes through point $(-3, 0)$. What is the equation of the line? _____

22) The slope of a line is 8 and it passes through point $(0, 7)$. What is the equation of the line? _____

WWW.MathNotion.Com

FSA Subject Test Mathematics Grade 7

Graphing Linear Inequalities

✏️ **Sketch the graph of each linear inequality.**

1) $y > 4x - 5$

2) $y < 2x + 4$

3) $y \leq -5x - 2$

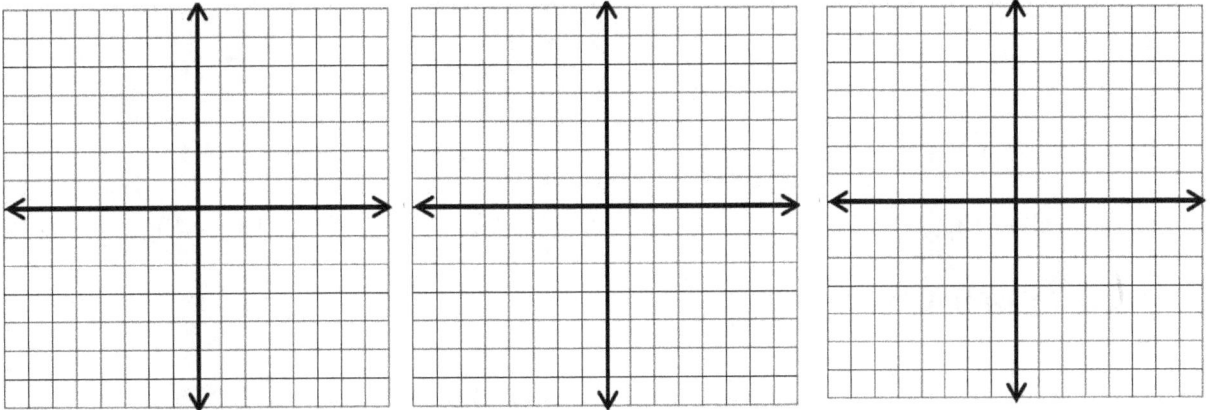

4) $4y \geq 12 + 4x$

5) $-12y < 3x - 24$

6) $5y \geq -15x + 10$

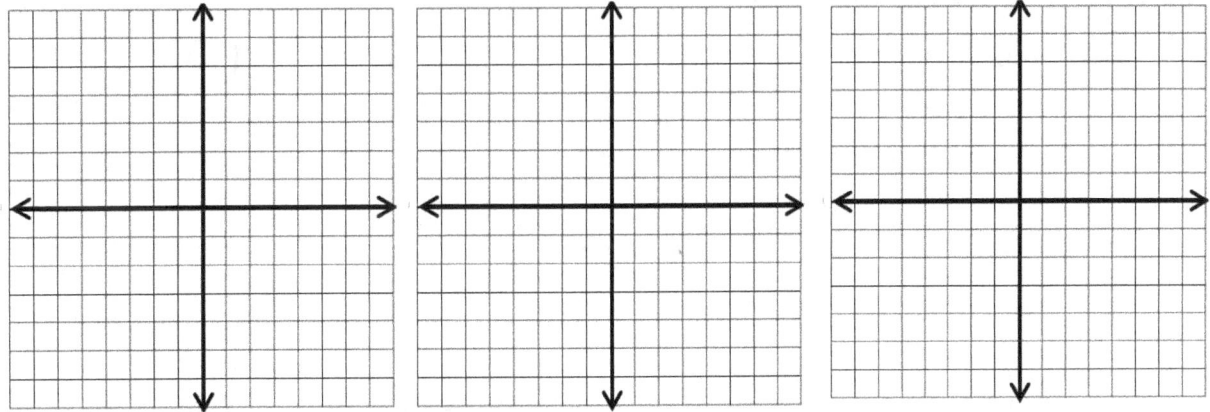

WWW.MathNotion.Com   87

FSA Subject Test Mathematics Grade 7

Finding Midpoint

✍ **Find the midpoint of the line segment with the given endpoints.**

1) $(-4, -3), (2, 3)$

2) $(9, 0), (-1, 8)$

3) $(9, -6), (3, 14)$

4) $(-10, -6), (0, 8)$

5) $(2, -5), (14, -15)$

6) $(-10, -3), (4, -13)$

7) $(8, 7), (-8, 13)$

8) $(-3, 6), (-9, 2)$

9) $(-4, 5), (16, -9)$

10) $(7, 14), (9, -2)$

11) $(-8, 6), (6, 6)$

12) $(10, 5), (-2, -3)$

13) $(-5, 12), (-3, 3)$

14) $(12, 7), (8, -2)$

15) $(10, 2), (-6, 14)$

16) $(-1, -2), (-7, 10)$

17) $(7, -7), (13, -13)$

18) $(-3, -8), (11, -4)$

19) $(5, -11), (-8, 9)$

20) $(14, -4), (16, 14)$

21) $(0, -5), (8, -1)$

22) $(3, 0), (-21, 18)$

23) $(17, -3), (-7, -5)$

24) $(26, -12), (6, 24)$

✍ **Find the answer for each problem.**

25) One endpoint of a line segment is $(-3, 7)$ and the midpoint of the line segment is $(-6, 9)$. What is the other endpoint? _____

26) One endpoint of a line segment is $(-3, 7)$ and the midpoint of the line segment is $(1, 5)$. What is the other endpoint? _____

27) One endpoint of a line segment is $(-10, -16)$ and the midpoint of the line segment is $(2, 9)$. What is the other endpoint? _____

WWW.MathNotion.Com

FSA Subject Test Mathematics Grade 7

Finding Distance of Two Points

✎ Find the distance between each pair of points.

1) $(6, 3), (-3, -9)$

2) $(5, 2), (-10, -6)$

3) $(8, 5), (8, 3)$

4) $(-8, -2), (2, 22)$

5) $(6, -7), (-3, -7)$

6) $(12, 0), (-9, -20)$

7) $(3, 20), (3, -5)$

8) $(10, 17), (5, 5)$

9) $(7, -2), (-4, -2)$

10) $(13, 4), (5, -2)$

11) $(11, 13), (5, 5)$

12) $(1, 4), (-23, -3)$

13) $(9, 8), (5, -4)$

14) $(-11, -4), (5, 8)$

15) $(-2, -6), (-2, -12)$

16) $(-1, -4), (23, 3)$

17) $(19, 3), (7, -6)$

18) $(-5, -2), (3, 4)$

19) $(2, 6), (2, -12)$

20) $(-4, -2), (8, -2)$

✎ Find the answer for each problem.

21) Triangle ABC is a right triangle on the coordinate system and its vertices are $(-2, 5), (-2, 1),$ and $(1, 1)$. What is the area of triangle ABC? _____

22) Three vertices of a triangle on a coordinate system are $(3, -6), (-5, -12),$ and $(3, -18)$. What is the perimeter of the triangle? _____

23) Four vertices of a rectangle on a coordinate system are $(-2, 2), (-2, 6), (4, 2),$ and $(4, 6)$. What is its perimeter? _____

WWW.MathNotion.Com

FSA Subject Test Mathematics Grade 7

Answers of Worksheets

Finding Slope

1) 1
2) −3
3) 2
4) −4
5) 6
6) −5
7) 8
8) −9
9) −7
10) 3
11) $\frac{1}{3}$
12) $-\frac{4}{5}$
13) $\frac{1}{2}$
14) −1
15) $\frac{1}{3}$
16) $\frac{1}{8}$
17) $\frac{7}{5}$
18) 2
19) −3
20) 2
21) −1
22) 2
23) 2
24) −3
25) −1
26) $\frac{5}{2}$
27) 2
28) −11

Graphing Lines Using Line Equation

1) $y = x - 2$

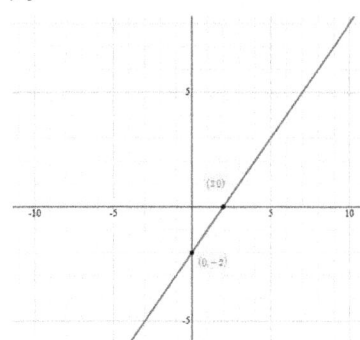

2) $y = -3x + 2$

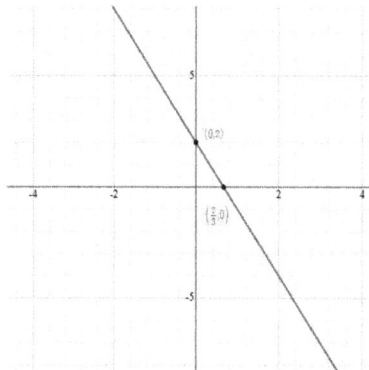

3) $x + y = 0$

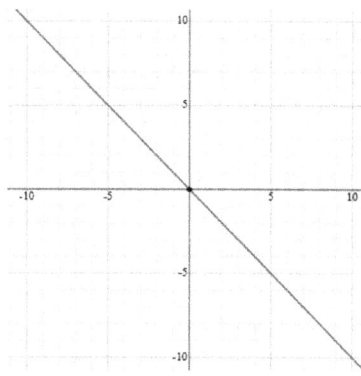

4) $x + y = -3$

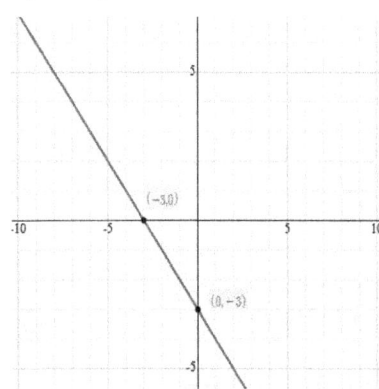

5) $2x + 3y = -4$

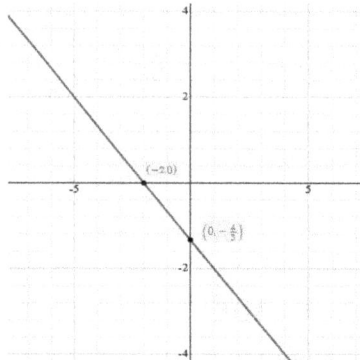

6) $y - 3x + 6 = 0$

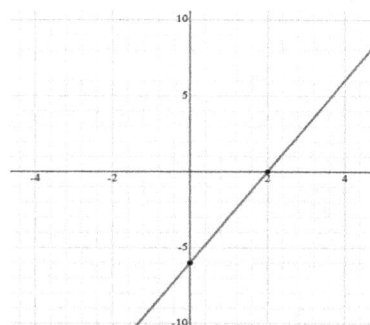

FSA Subject Test Mathematics Grade 7

Writing Linear Equations

1) $y = 14x - 33$
2) $y = x + 9$
3) $y = -4x + 47$
4) $y = x - 4$
5) $y = 2x - 31$
6) $y = 9x - 43$
7) $y = -2x + 24$
8) $y = -3x + 1$
9) $y = -2x + 5$
10) $y = -2x + 6$
11) $y = 5x - 18$
12) $y = 14x - 122$
13) $y = 13x + 52$
14) $y = x + 10$
15) $y = -5x + 58$
16) $y = -5x + 5$
17) $y = 6x + 12$
18) $y = -11x - 4$
19) $y = -3x + 17$
20) $y = -5x - 11$
21) $y = -10x - 30$
22) $y = 8x + 7$

Graphing Linear Inequalities

1) $y > 4x - 5$
2) $y < 2x + 4$
3) $y \le -5x - 2$

4) $4y \ge 12 + 4x$

5) $-12y < 3x - 24$
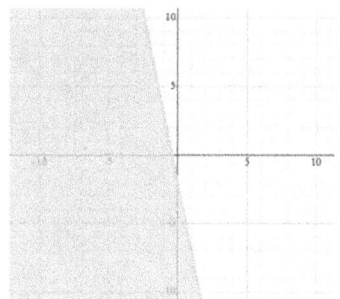

6) $5y \ge -15x + 10$
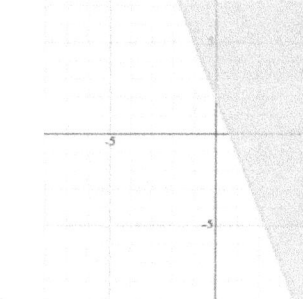

Finding Midpoint

1) $(-1, 0)$
2) $(4, 4)$
3) $(6, 4)$
4) $(-5, 1)$
5) $(8, -10)$
6) $(-3, -8)$
7) $(0, 10)$
8) $(-6, 4)$
9) $(6, -2)$
10) $(8, 6)$
11) $(-1, 6)$
12) $(4, 1)$
13) $(-4, 7.5)$
14) $(10, 2.5)$
15) $(2, 8)$
16) $(-4, 4)$
17) $(10, -10)$
18) $(4, -6)$

WWW.MathNotion.Com

FSA Subject Test Mathematics Grade 7

19) $(-1.5, -1)$
20) $(15, 5)$
21) $(4, -3)$

22) $(-9, 9)$
23) $(5, -4)$
24) $(16, 6)$

25) $(-9, 11)$
26) $(5, 3)$
27) $(14, 34)$

Finding Distance of Two Points

1) 15
2) 17
3) 2
4) 26
5) 9
6) 29
7) 25
8) 13

9) 11
10) 10
11) 10
12) 25
13) $4\sqrt{10}$
14) 20
15) 6
16) 25

17) 15
18) 10
19) 18
20) 12
21) 6 *square units*
22) 32 *units*
23) 20 *units*

FSA Subject Test Mathematics Grade 7

Chapter 9 :
Transformations

Topics that you will learn in this chapter:

- ✓ Translations
- ✓ Reflections
- ✓ Rotations
- ✓ Dilations
- ✓ Coordinate of Vertices.

"Nothing gets transformed in your life until your mind is transformed."

Ifeanyi Enoch Onuoha

FSA Subject Test Mathematics Grade 7

Translations

✎ **Graph the image of the figure using the transformation given.**

1) Translation: 5 units right and 2 units down

2) Translation: 1 units left and 3 units down

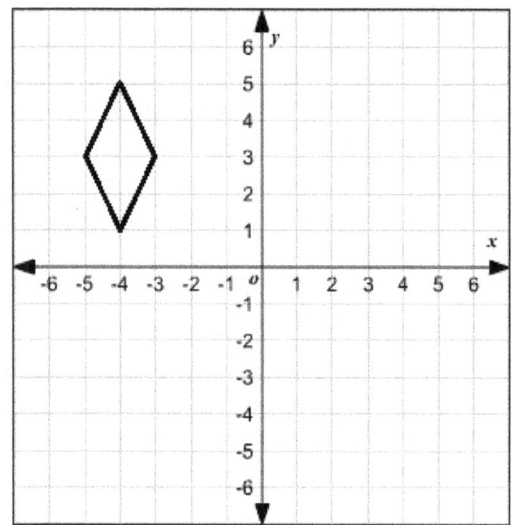
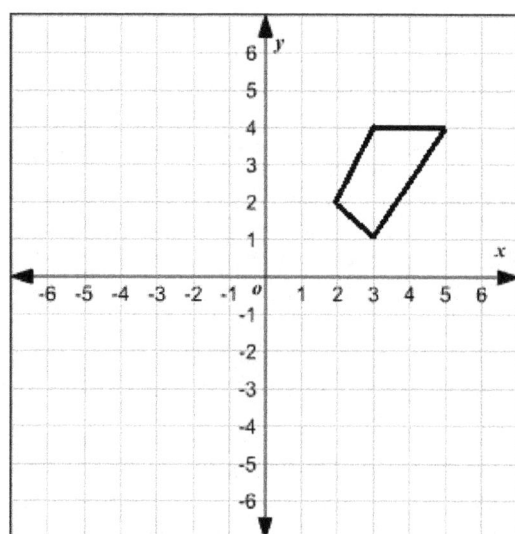

✎ **Write a rule to describe each transformation.**

3)

4)

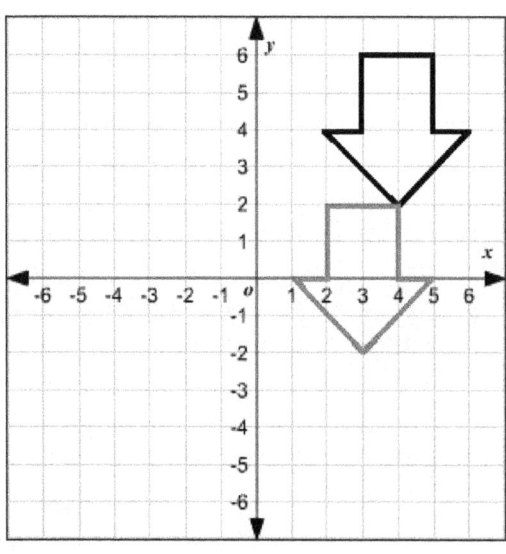
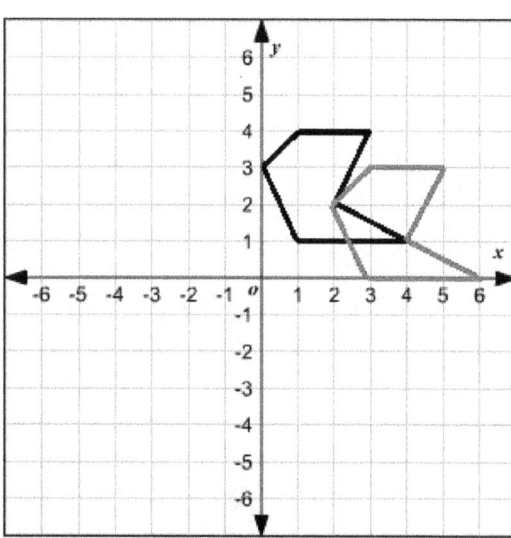

FSA Subject Test Mathematics Grade 7

Reflections

✏️ **Graph the image of the figure using the transformation given.**

1) Reflection across $x = 3$

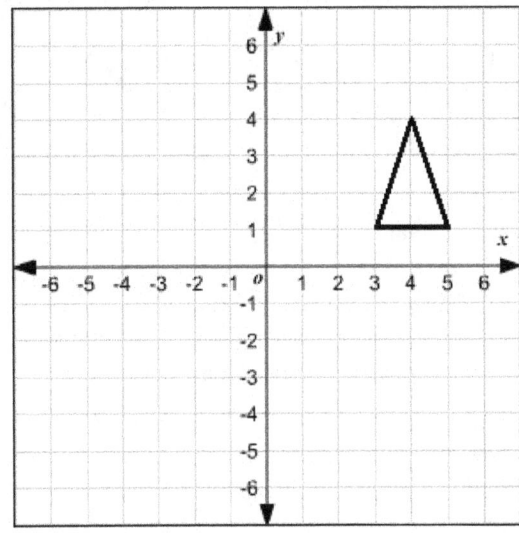

2) Reflection across $y = x$

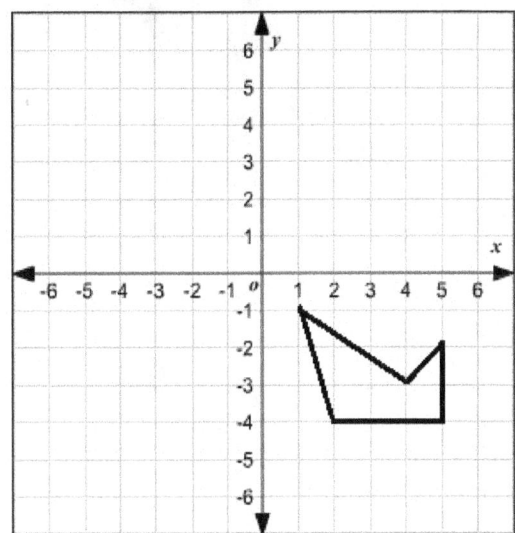

3) Reflection across $y = -2$

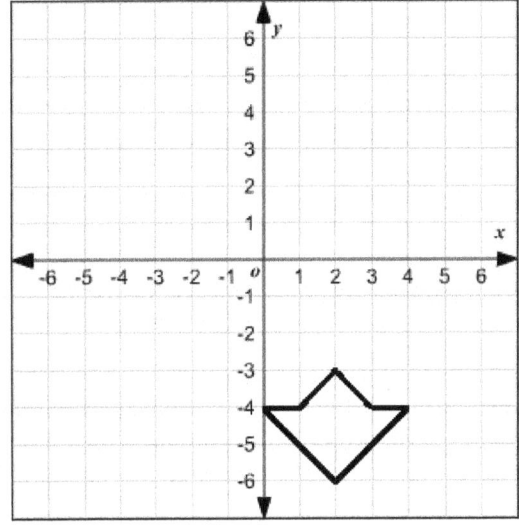

4) Reflection across y axis

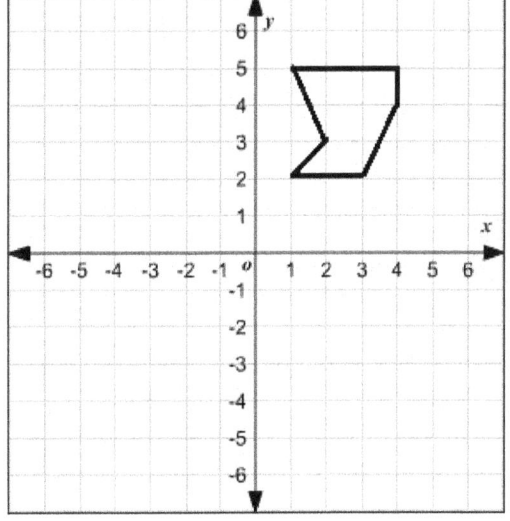

WWW.MathNotion.Com

FSA Subject Test Mathematics Grade 7

✎ **Write a rule to describe each transformation.**

5)

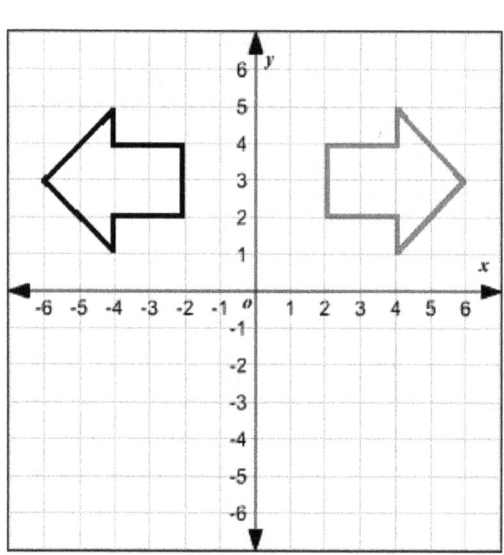

6)

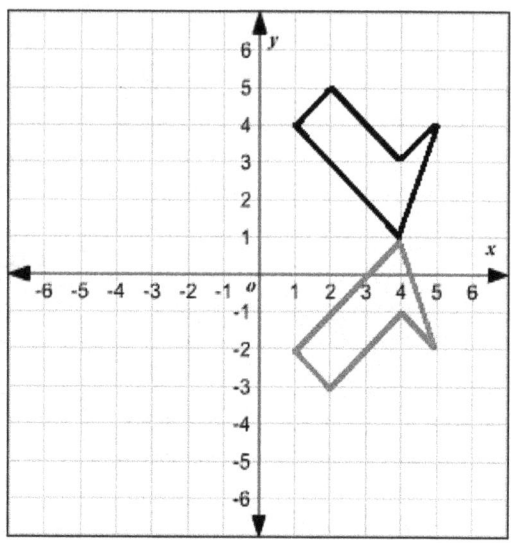

7)

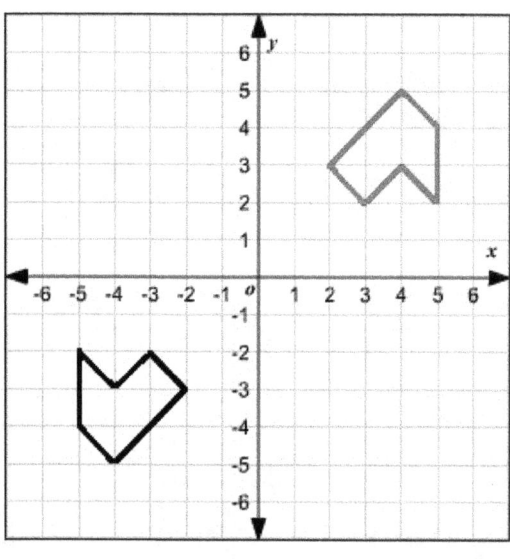

8)

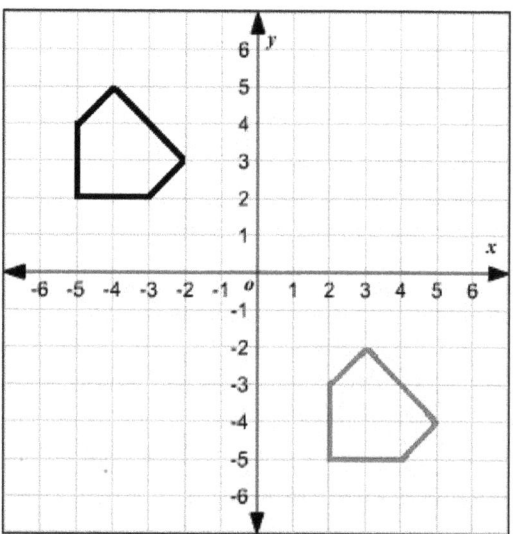

FSA Subject Test Mathematics Grade 7

Rotations

✎ **Graph the image of the figure using the transformation given.**

1) Rotation 90° counterclockwise about the origin

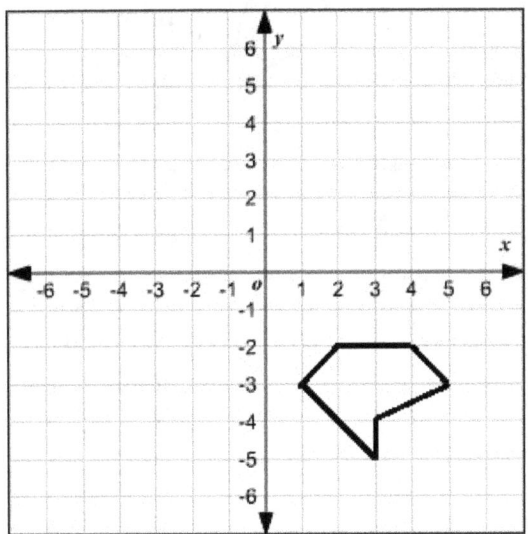

2) Rotation 270° counterclockwise about the origin

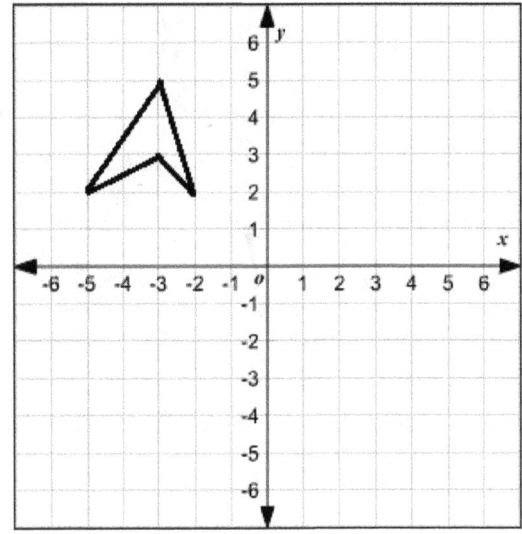

3) Rotation 270° clockwise about the origin

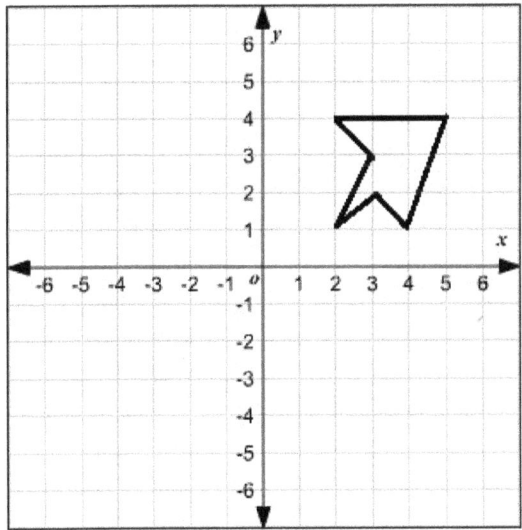

4) Rotation 180° counterclockwise about the origin

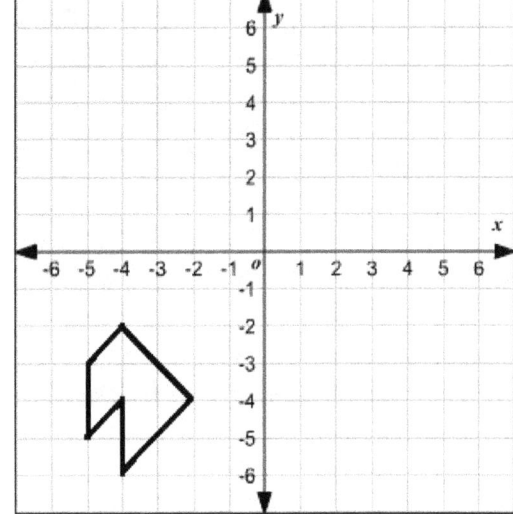

WWW.MathNotion.Com

FSA Subject Test Mathematics Grade 7

✎ **Write a rule to describe each transformation.**

5)

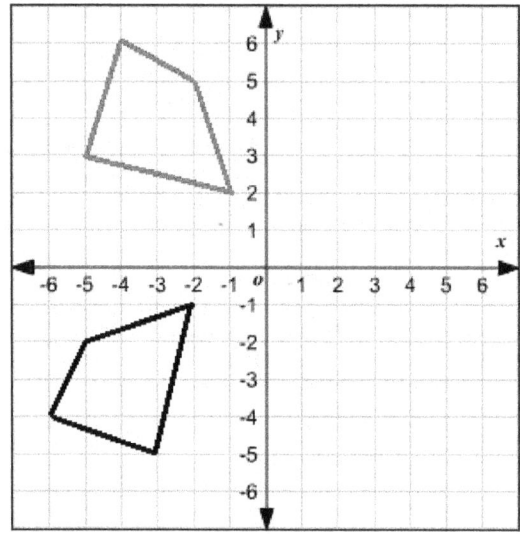

6)

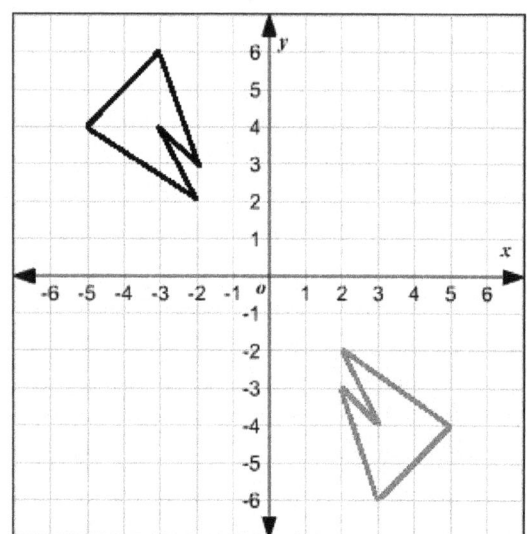

7)

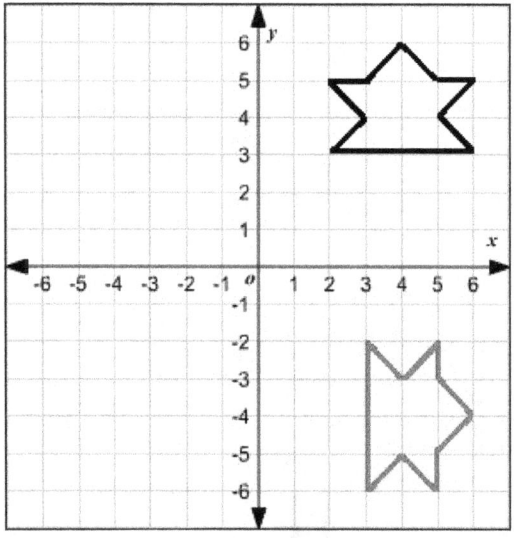

8)

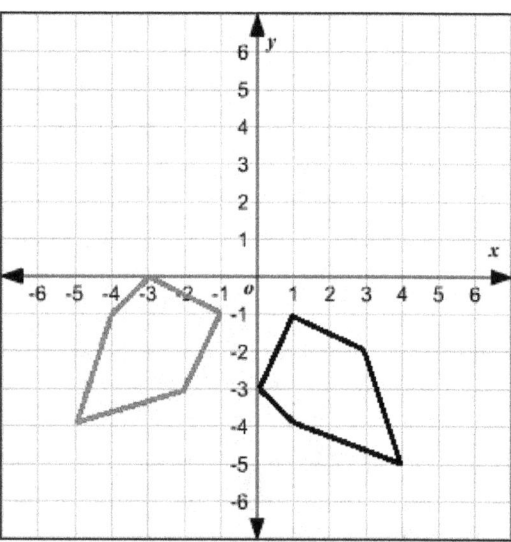

WWW.MathNotion.Com

FSA Subject Test Mathematics Grade 7

Dilations

✎ **Determine whether the dilation from figure M to figure N is a reduction or an enlargement. Then find the scale factor and the missing length.**

1)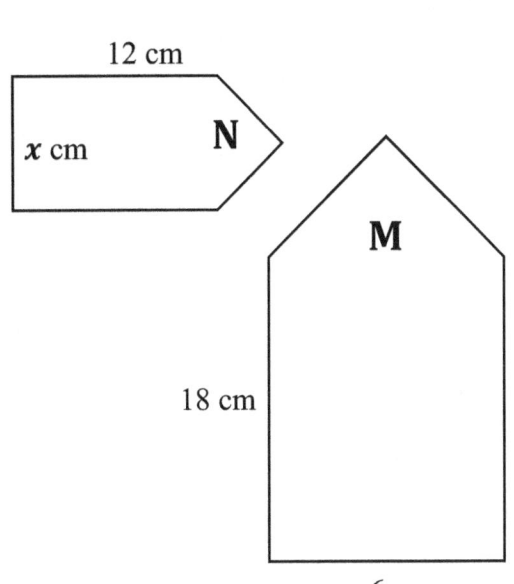

2)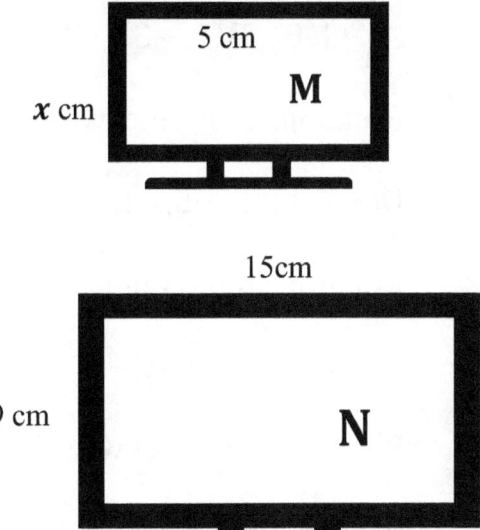

✎ **Draw a dilation of the figure using the given scale factor.**

3) $k = \frac{1}{2}$

4) $k = 3$

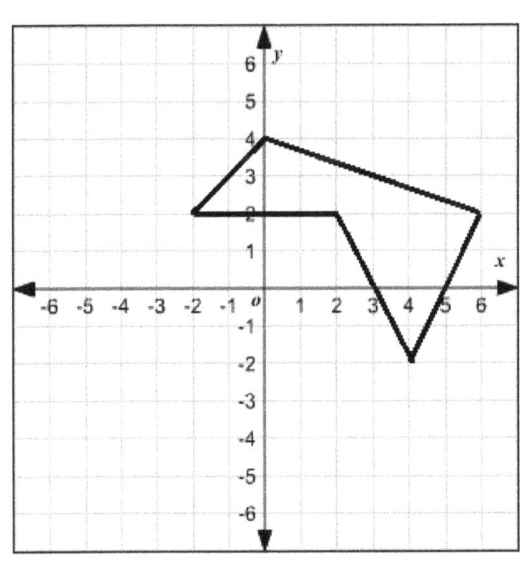

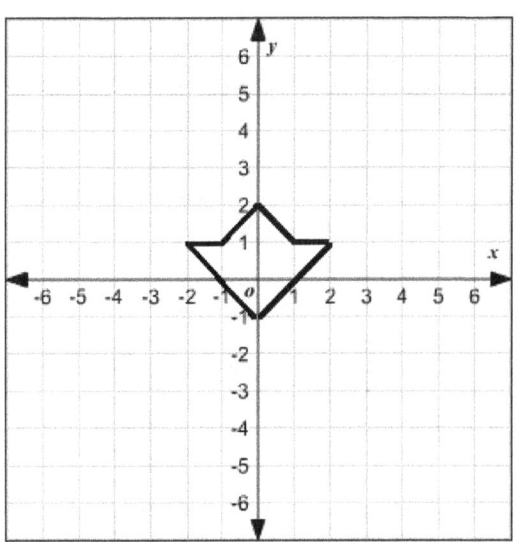

www.MathNotion.Com

FSA Subject Test Mathematics Grade 7

Coordinates of Vertices

✎ **Calculate the new coordinates after the given transformations.**

1) Translate: 1-unit right and 2 units down.

   A(−1, 0), B(1, −3), C(2, 1)

2) Rotation: 270° clockwise about the origin.

   D(−1, 1), E(−5, 6), F(−7, 3), G(−3, 10)

3) Rotation: 180° counterclockwise about the origin.

   P(3, 0), Q(5, 2), R(−1, 4), S(2, −7)

4) Rotation: 90° clockwise about the origin.

   J(−2, 5), K(−5, 0), L(3, −6)

5) Reflection: over the y axis.

   C(−2, −6), D(4, −3), W(1, −7), Y(5, 1)

6) Reflection: across the line $y = -x$.

   A(3, −2), B(8, −4), C(6, −6), D(1, −5)

7) Reflection: across the line $y = -2$.

   K(−1, 1), L(−4, 2), M(4, −1), N(2, 3)

8) Dilate: Reduction by scale factor $\frac{1}{3}$.

   A(6, 3), B(−12, 0), C(−9, 6)

9) Dilate: Enlargement by scale factor 2.

   F(−1, 4), G(−3, 0), H(3, 2)

WWW.MathNotion.Com

FSA Subject Test Mathematics Grade 7

Answers of Worksheets

Translations

1)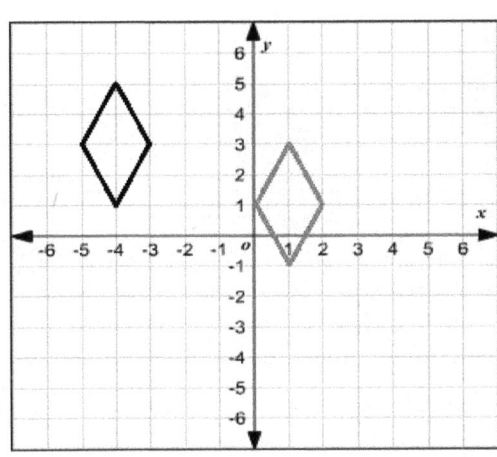

2)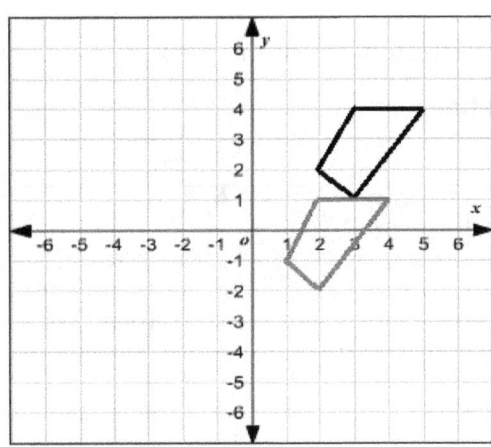

3) Translation: : 1 units left and 4 units down

4) Translation: 2 units right and 1 unit down

Reflections

1)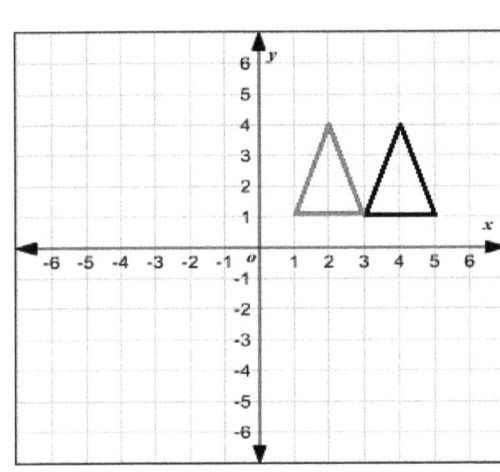

2)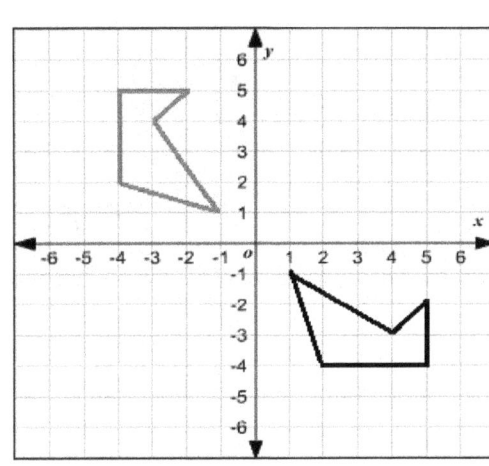

3)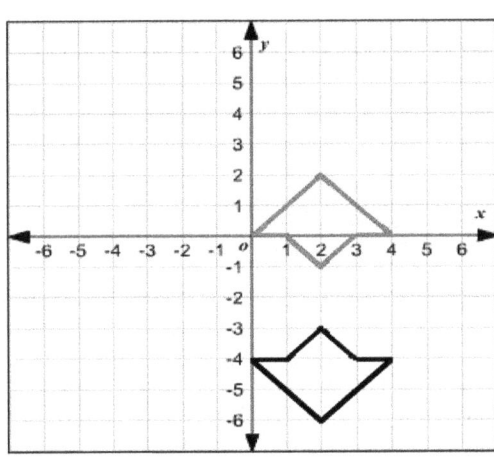

4)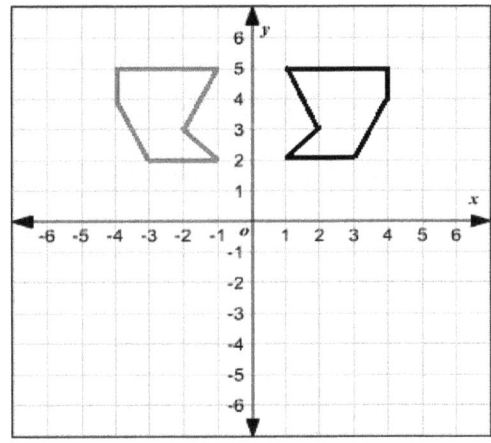

WWW.MathNotion.Com

FSA Subject Test Mathematics Grade 7

5) Reflection across the x = 0 (y axis)

6) Reflection across the y = 1

7) Reflection against the origin

8) Reflection across the y = x

Rotations

1)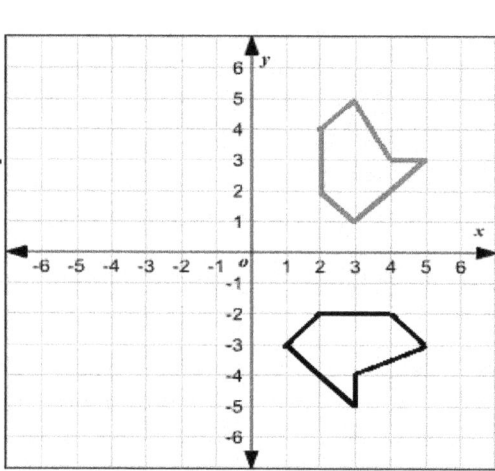

2)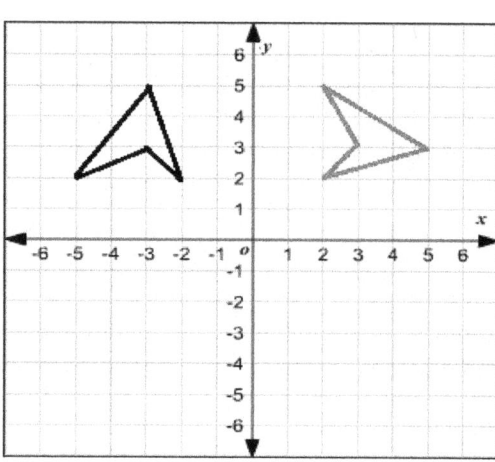

3)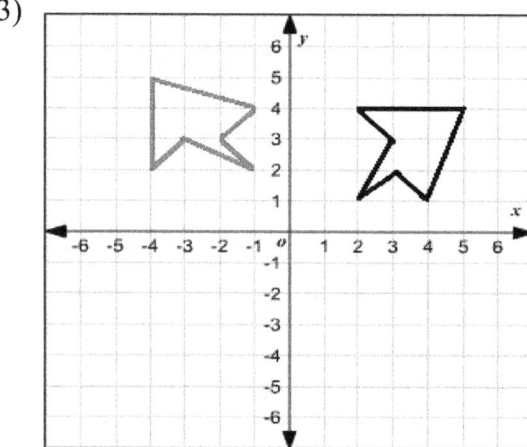

4)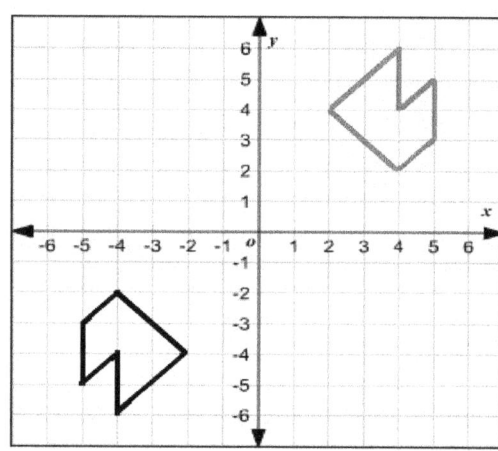

5) Rotation 90° clockwise about the origin

6) Rotation 180° about the origin

7) Rotation 270° counter clockwise about the origin

8) Rotation 90° clockwise about the origin

Dilations

1) Reduction, $k = \frac{3}{2}$, $x = 4\ cm$

2) Enlargement, $k = \frac{1}{3}$, $x = 3\ cm$

FSA Subject Test Mathematics Grade 7

3)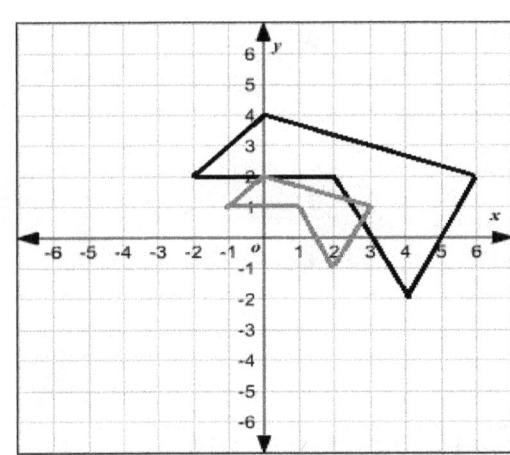

4)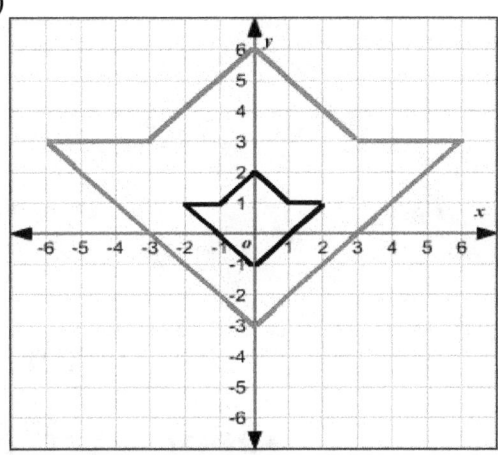

Coordinate of Vertices

1) $A'(0,-2), B'(2,-5), C'(3,-1)$
2) $D'(-1,-1), E'(-6,-5), F'(-3,-7), G'(-10,-3)$
3) $P'(-3,0), Q'(-5,-2), R'(1,-4), S'(-2,7)$
4) $J'(5,2), K'(0,5), L'(-6,-3)$
5) $C'(2,-6), D'(-4,-3), W'(-1,-7), Y'(-5,1)$
6) $A'(2,-3), B'(4,-8), C'(6,-6), D'(5,-1)$
7) $K'(-1,-5), L'(-4,-6), M'(4,-3), N'(2,-7)$
8) $A'(2,1), B'(-4,0), C'(-3,2)$
9) $F'(-2,8), G'(-6,0), H'(6,4)$

FSA Subject Test Mathematics Grade 7

Chapter 10:
Geometry and Solid Figures

Topics that you will practice in this chapter:

- ✓ Angles
- ✓ Pythagorean Relationship
- ✓ Triangles
- ✓ Polygons
- ✓ Trapezoids
- ✓ Circles
- ✓ Cubes
- ✓ Rectangular Prism
- ✓ Cylinder
- ✓ Pyramids and Cone

Mathematics is, as it were, a sensuous logic, and relates to philosophy as do the arts, music, and plastic art to poetry. — K. Shegel

FSA Subject Test Mathematics Grade 7

Angles

✏️ **What is the value of x in the following figures?**

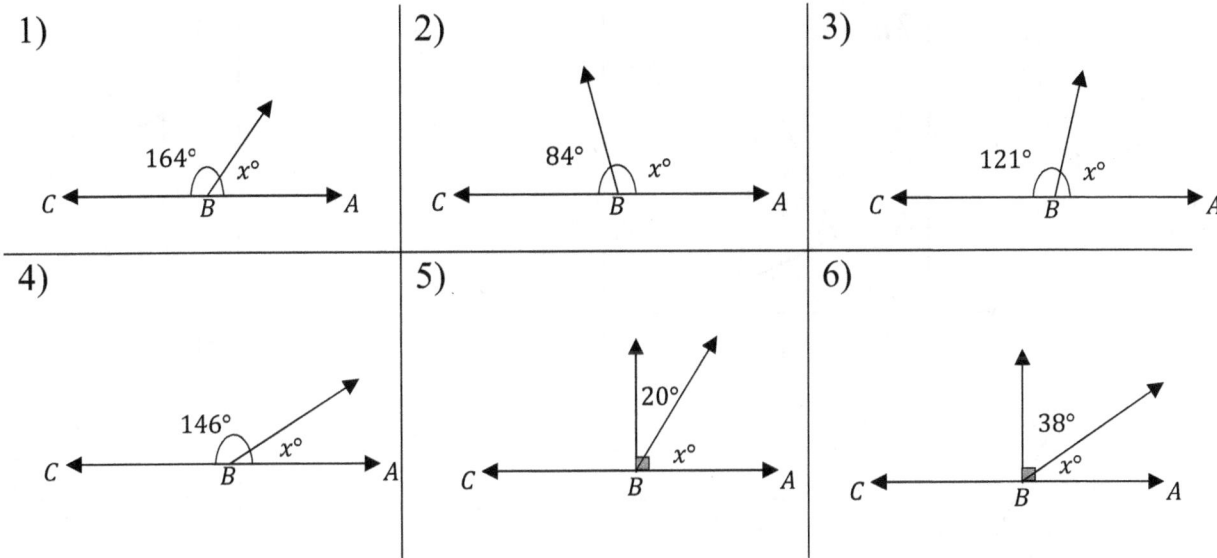

1) 164°, $x°$
2) 84°, $x°$
3) 121°, $x°$
4) 146°, $x°$
5) 20°, $x°$
6) 38°, $x°$

✏️ **Calculate.**

7) Two supplement angles have equal measures. What is the measure of each angle? _____

8) The measure of an angle is seven fifth the measure of its supplement. What is the measure of the angle? _____

9) Two angles are complementary and the measure of one angle is 24 less than the other. What is the measure of the smaller angle? _____

10) Two angles are complementary. The measure of one angle is one fifth the measure of the other. What is the measure of the bigger angle? _____

11) Two supplementary angles are given. The measure of one angle is 40° less than the measure of the other. What does the smaller angle measure? _____

FSA Subject Test Mathematics Grade 7

Pythagorean Relationship

✎ Do the following lengths form a right triangle?

1)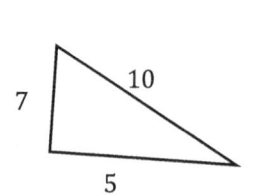

2)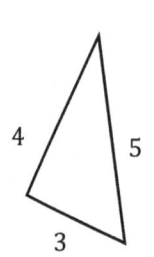

3)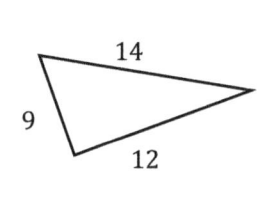

4)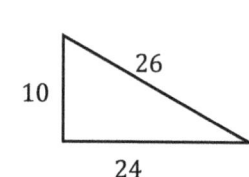

5)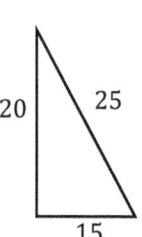

6)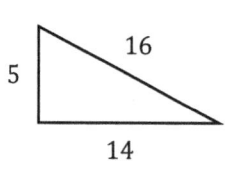

7)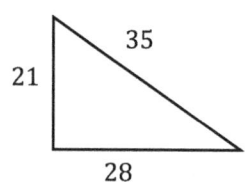

8)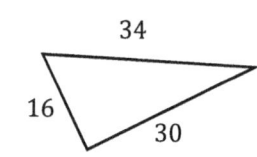

✎ Find the missing side?

9)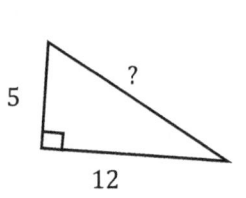

10)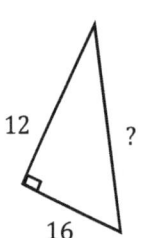

11)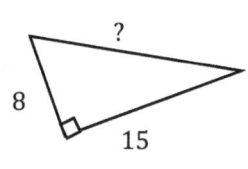

12)

13)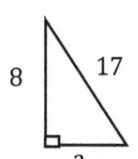

14)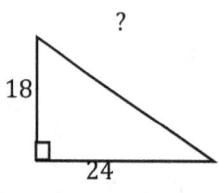

15)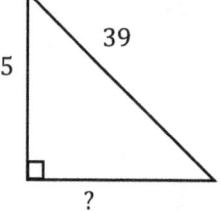

16)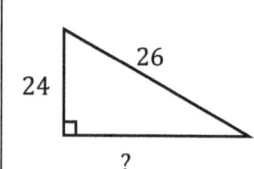

WWW.MathNotion.Com

FSA Subject Test Mathematics Grade 7

Triangles

✎ Find the measure of the unknown angle in each triangle.

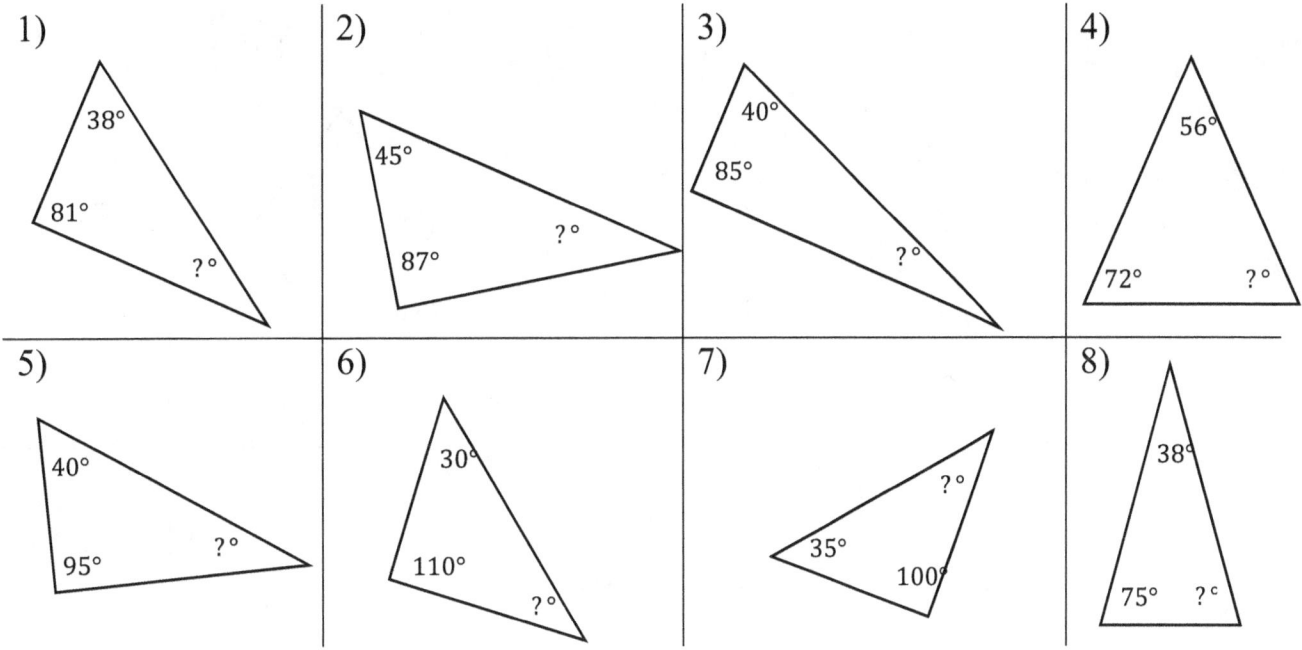

✎ Find area of each triangle.

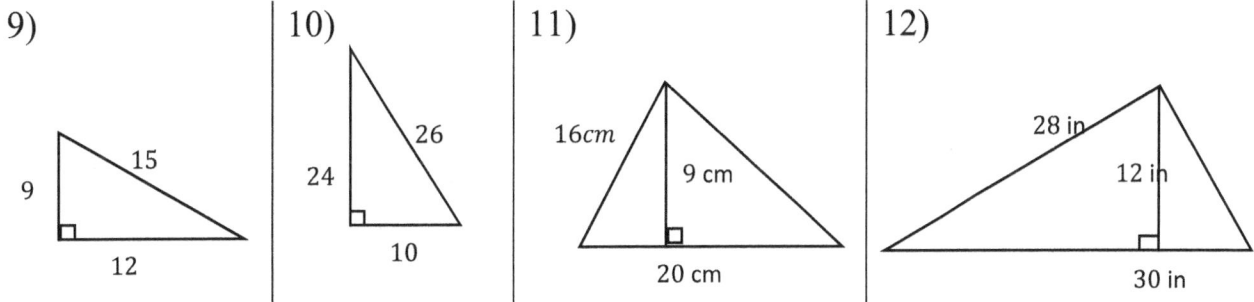

WWW.MathNotion.Com

FSA Subject Test Mathematics Grade 7

Polygons

✏️ **Find the perimeter of each shape.**

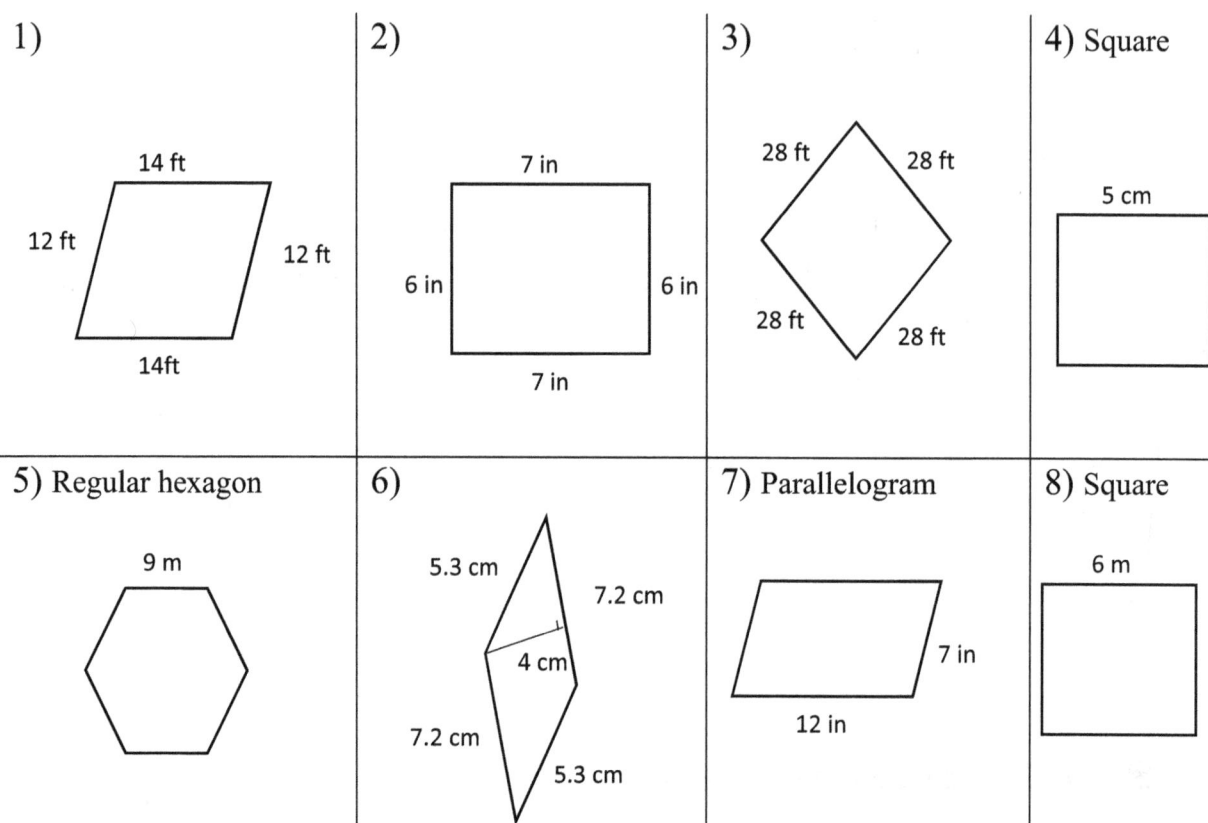

✏️ **Find the area of each shape.**

9) Parallelogram
5 m
6 m
5 m

10) Rectangle
25 in
12 in

11) Rectangle
16 km
10 km

12) Square
7 in

WWW.MathNotion.Com   108

FSA Subject Test Mathematics Grade 7

Trapezoids

🖎 Find the area of each trapezoid.

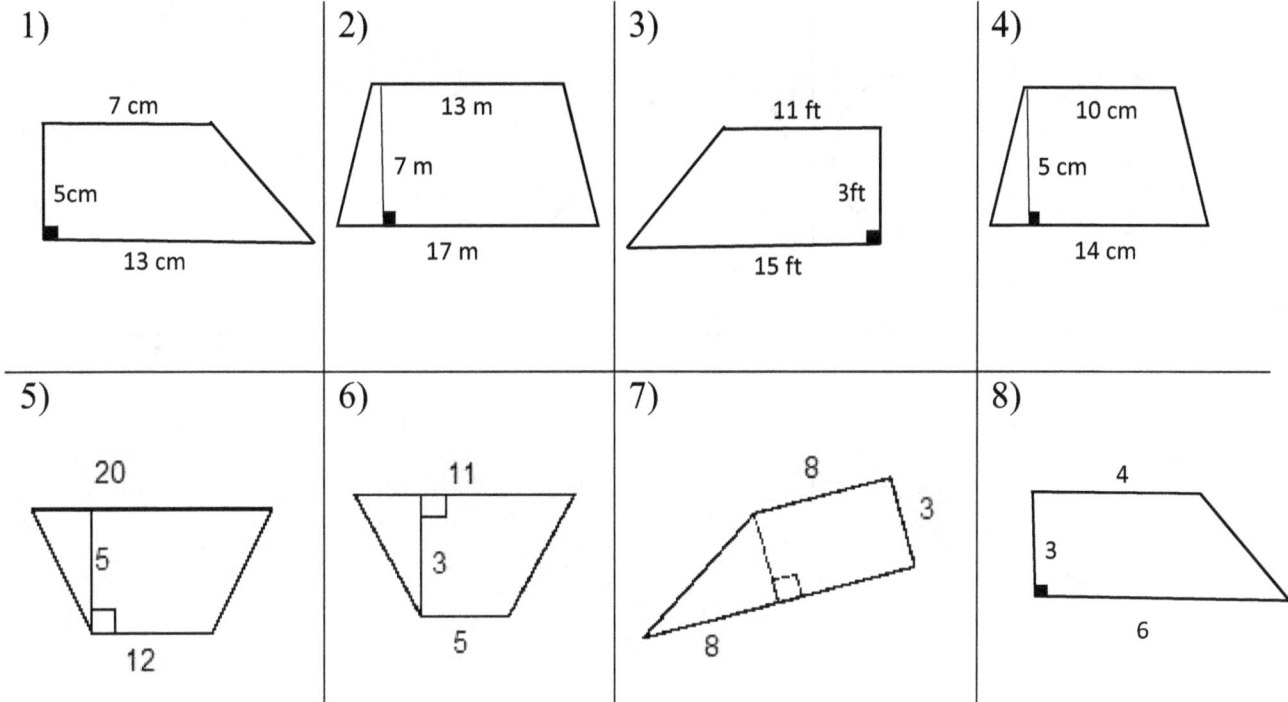

🖎 Calculate.

1) A trapezoid has an area of 45 cm² and its height is 5 cm and one base is 5 cm. What is the other base length? _____

2) If a trapezoid has an area of 99 ft² and the lengths of the bases are 8 ft and 10 ft, find the height? _____

3) If a trapezoid has an area of 126 m² and its height is 14 m and one base is 6 m, find the other base length? _____

4) The area of a trapezoid is 440 ft² and its height is 22 ft. If one base of the trapezoid is 15 ft, what is the other base length? _____

FSA Subject Test Mathematics Grade 7

Circles

✏️ **Find the area of each circle.** ($\pi = 3.14$)

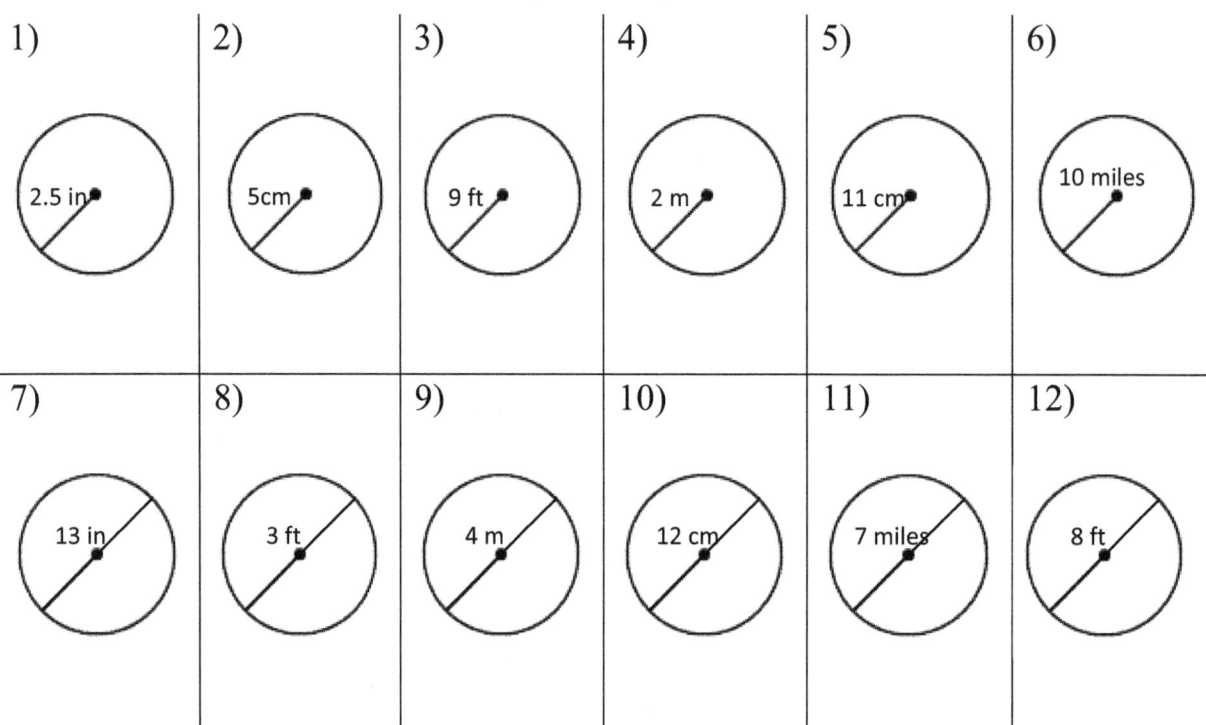

✏️ **Complete the table below.** ($\pi = 3.14$)

Circle No.	Radius	Diameter	Circumference	Area
1	1 in	2 in	6.28 in	3.14 in^2
2		10 m		
3				28.26 ft^2
4			47.1 mi	
5		11 km		
6	7 cm			
7		12 ft		
8				314 m^2
9			56.52 in	
10	4.5 ft			

WWW.MathNotion.Com

FSA Subject Test Mathematics Grade 7

Cubes

✏ Find the volume of each cube.

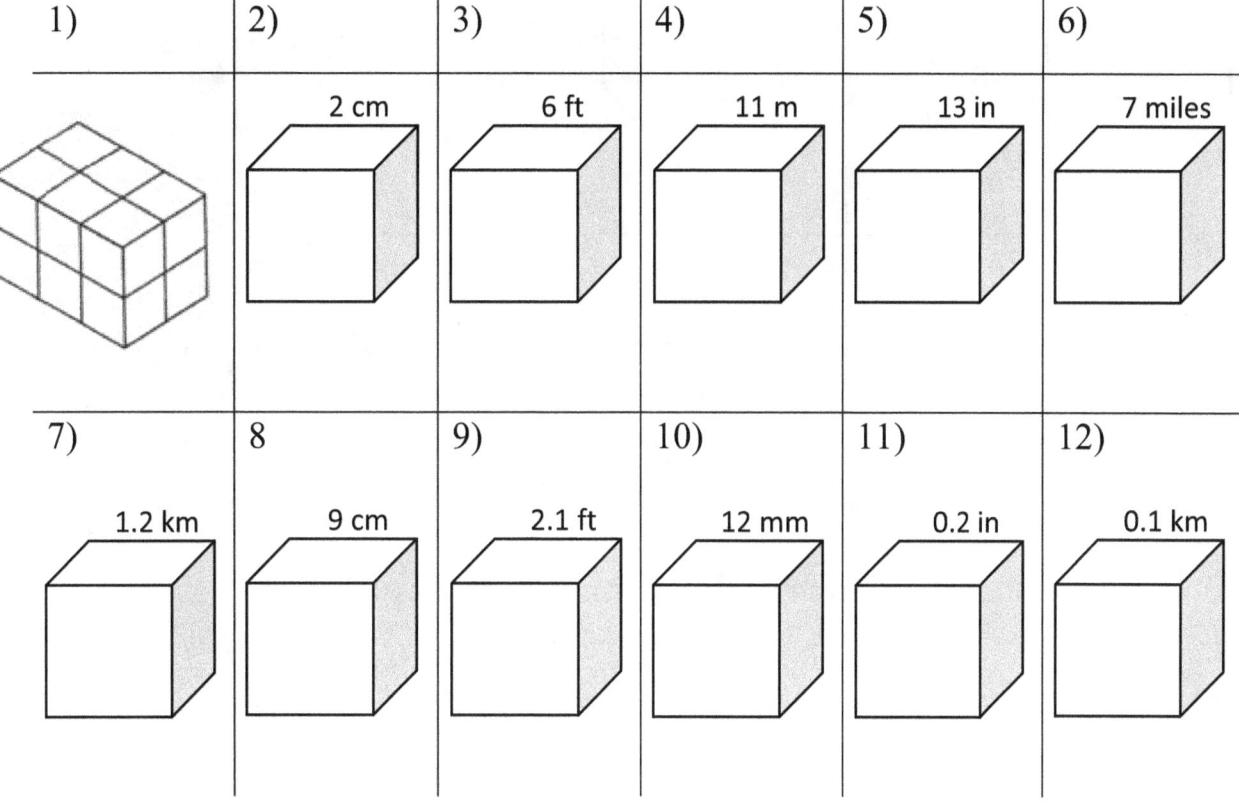

✏ Find the surface area of each cube.

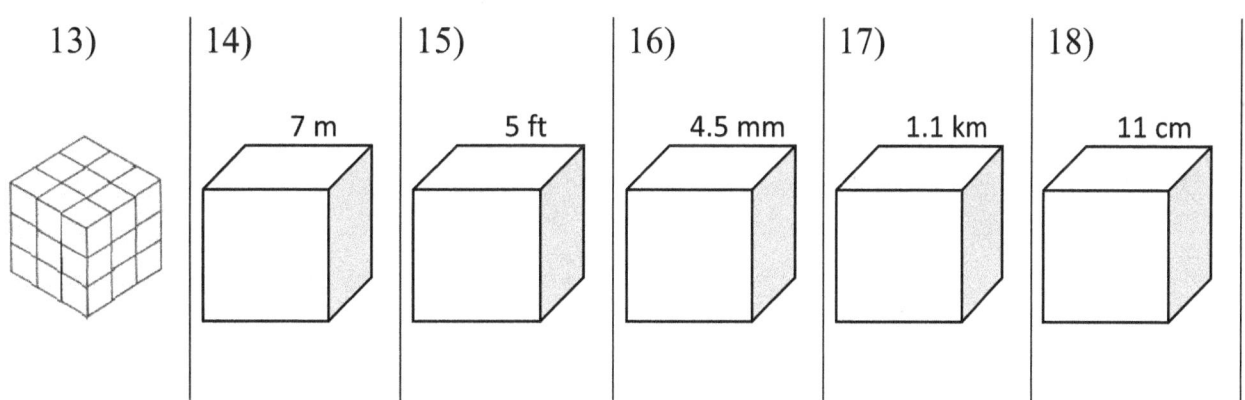

Rectangular Prism

✏ Find the volume of each Rectangular Prism.

FSA Subject Test Mathematics Grade 7

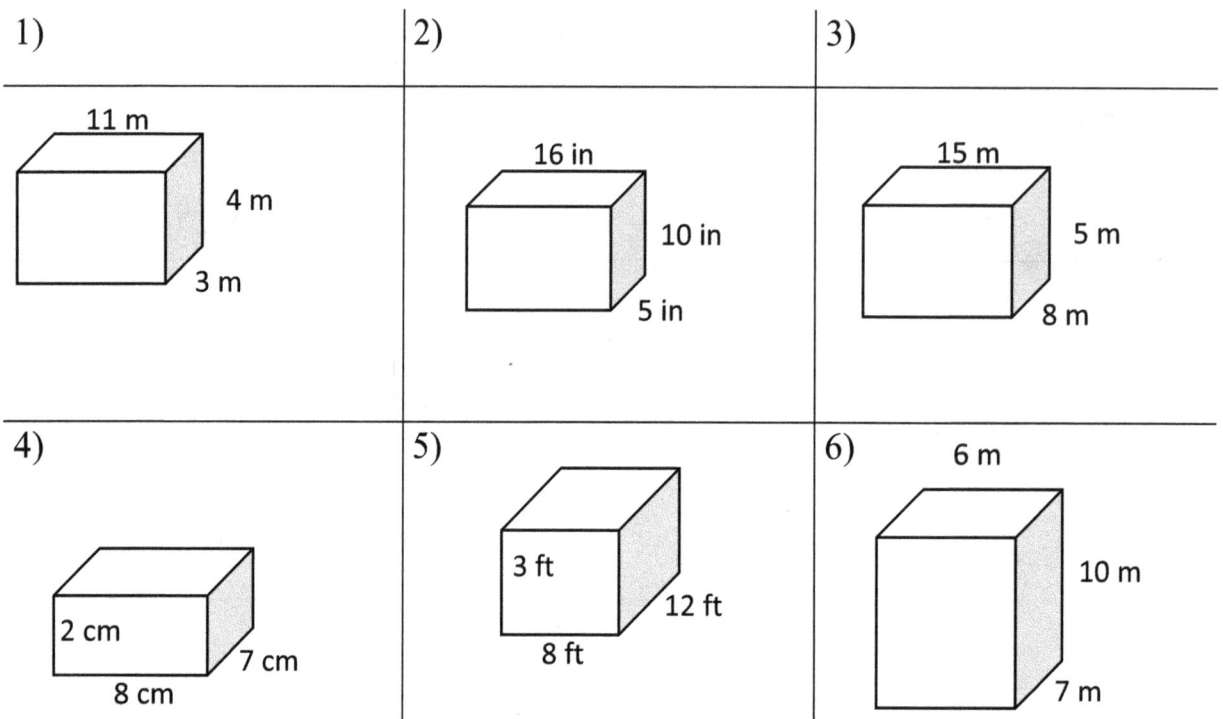

🖎 **Find the surface area of each Rectangular Prism.**

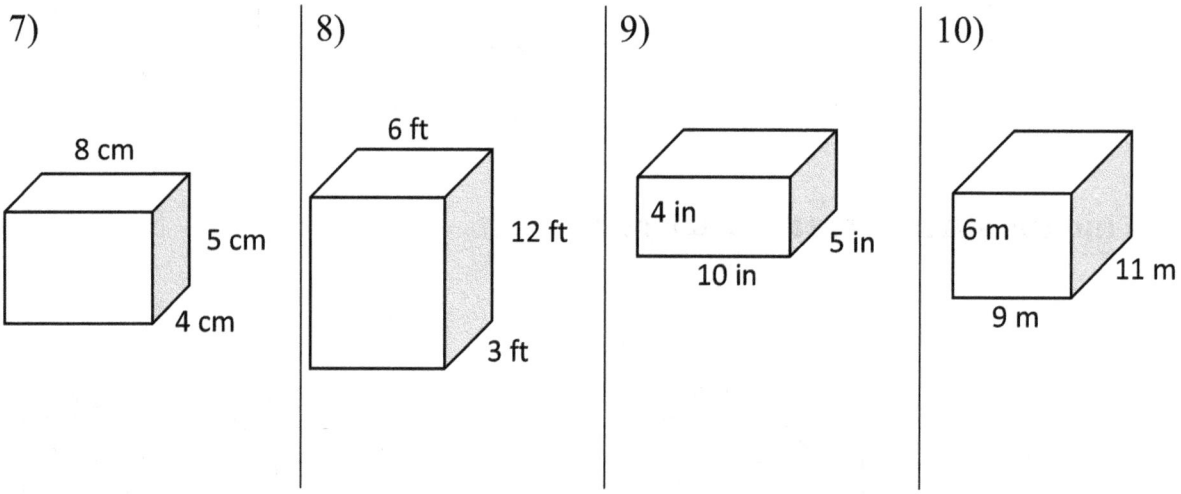

WWW.MathNotion.Com

FSA Subject Test Mathematics Grade 7

Cylinder

✎ **Find the volume of each Cylinder. Round your answer to the nearest tenth.** ($\pi = 3.14$)

1)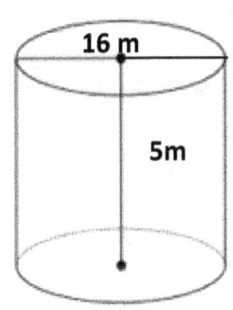

2)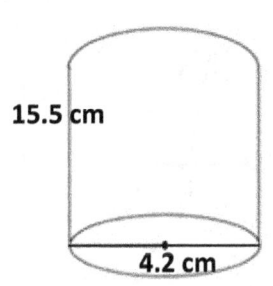

3)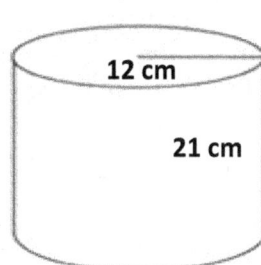

4)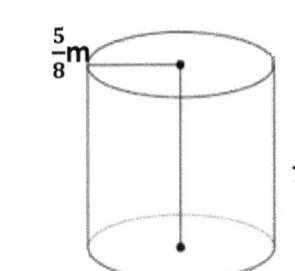

5)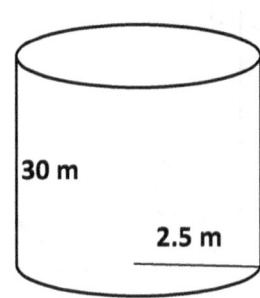

6)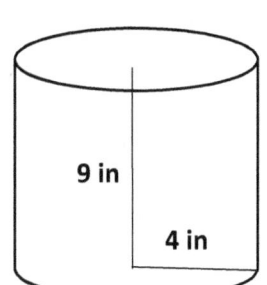

✎ **Find the surface area of each Cylinder.** ($\pi = 3.14$)

7)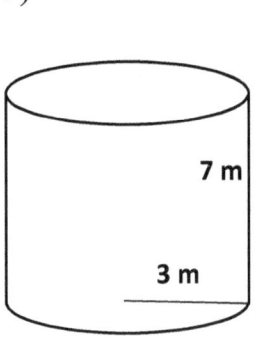

8)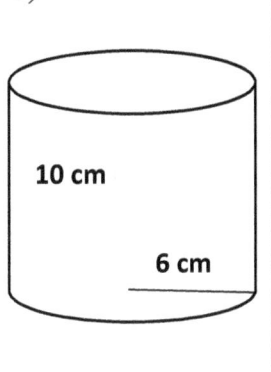

9)

10)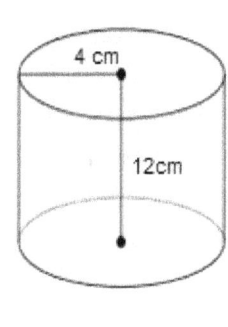

FSA Subject Test Mathematics Grade 7

Pyramids and Cone

✎ Find the volume of each Pyramid and Cone. ($\pi = 3.14$)

1)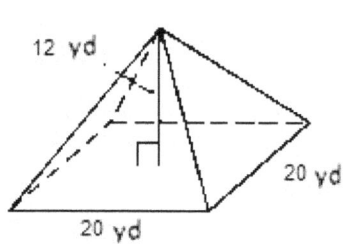

2)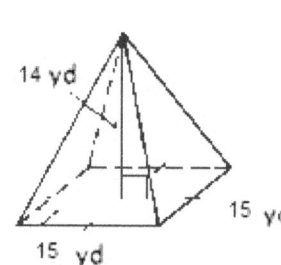

3)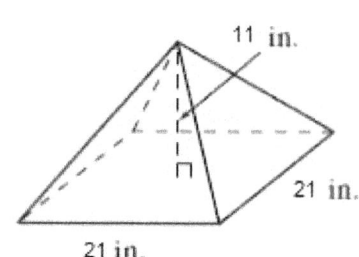

4)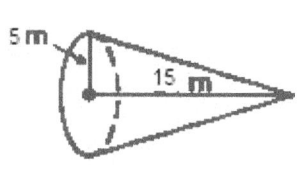

5)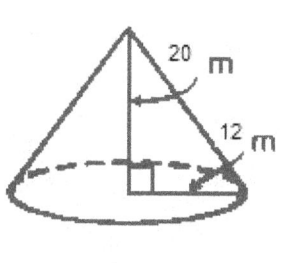

6)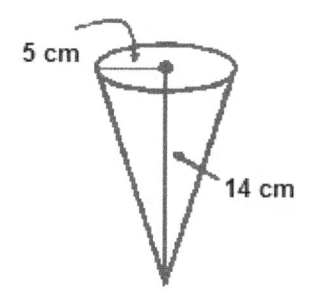

✎ Find the surface area of each Pyramid and Cone. ($\pi = 3.14$)

7)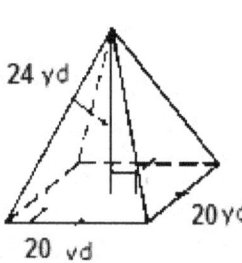

8)

9)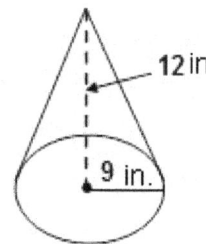

10)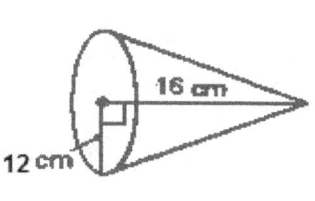

WWW.MathNotion.Com

FSA Subject Test Mathematics Grade 7

Answers of Worksheets

Angles

1) 16°
2) 96°
3) 59°
4) 34°
5) 70°
6) 52°
7) 90°
8) 75°
9) 33°
10) 75°
11) 70°

Pythagorean Relationship

1) No
2) Yes
3) No
4) Yes
5) Yes
6) No
7) Yes
8) Yes
9) 13
10) 20
11) 17
12) 10
13) 15
14) 30
15) 36
16) 12

Triangles

1) 60°
2) 48°
3) 55°
4) 52°
5) 45°
6) 40°
7) 45°
8) 67°
9) 54 square unites
10) 120 square unites
11) 90 square unites
12) 180 square unites

Polygons

1) 52 ft
2) 26 in
3) 112 ft
4) 20 cm
5) 54 m
6) 25 cm
7) 38 in
8) 24 m
9) 30 m^2
10) 300 in^2
11) 160 km^2
12) 49 in^2

Trapezoids

1) 50 cm^2
2) 105 m^2
3) 39 ft^2
4) 60 cm^2
5) 80
6) 24
7) 36
8) 15

Calculate

1) 13 cm
2) 11 ft
3) 12 m
4) 25 ft

Circles

1) 19.63 in^2
2) 78.5 cm^2
3) 254.34 ft^2
4) 12.56 m^2
5) 379.94 cm^2
6) 314 $miles^2$
7) 132.67 in^2
8) 7.07 ft^2
9) 12.56 m^2
10) 113.04 cm^2
11) 38.47 $miles^2$
12) 50.24 ft^2

FSA Subject Test Mathematics Grade 7

Circle No.	Radius	Diameter	Circumference	Area
1	1 in	2 in	6.28 in	3.14 in^2
2	5 m	10 m	31.4 m	78.5 m^2
3	3 ft	6 ft	18.84 ft	28.26 ft^2
4	7.5 miles	15 mi	47.1 mi	176.63 mi^2
5	5.5 km	11 km	34.54 km	94.99 km^2
6	7 cm	14 cm	43.96 cm	153.86 cm^2
7	6 ft	12 ft	37.68 feet	113.04 ft^2
8	10 m	20 m	62.8 m	314 m^2
9	9 in	18 in	56.52 in	254.34 in^2
10	4.5 ft	9 ft	28.26 ft	63.585 ft^2

Cubes

1) 12
2) 8 cm^3
3) 216 ft^3
4) 1,331 m^3
5) 2,197 in^3
6) 343 $miles^3$
7) 1.728 km^3
8) 729 cm^3
9) 9.261 ft^3
10) 1,728 mm^3
11) 0.008 in^3
12) 0.001 km^3
13) 27
14) 294 m^2
15) 150 ft^2
16) 121.5 mm^2
17) 7.26 km^2
18) 726 cm^2

Rectangular Prism

1) 132 m^3
2) 800 in^3
3) 600 m^3
4) 112 cm^3
5) 288 ft^3
6) 420 m^3
7) 184 cm^2
8) 252 ft^2
9) 220 in^2
10) 438 m^2

Cylinder

1) 1,004.8 m^3
2) 214.6 cm^3
3) 9,495.4 cm^3
4) 1.1 m^3
5) 588.8 m^3
6) 452.2 in^3
7) 188.4 m^2
8) 602.9 cm^2
9) 37.7 cm^2
10) 401.9 m^2

Pyramids and Cone

1) 1,600 yd^3
2) 1,050 yd^3
3) 1,617 in^3
4) 392.5 m^3
5) 3,014.4 m^3
6) 366.33 cm^3
7) 1,440 yd^2
8) 1,536 m^2
9) 678.24 in^2
10) 1,205.76 cm^2

FSA Subject Test Mathematics Grade 7

Chapter 11:
Statistics and Probability

Topics that you will practice in this chapter:

- ✓ Mean and Median
- ✓ Mode and Range
- ✓ Histograms
- ✓ Stem–and–Leaf Plot
- ✓ Pie Graph
- ✓ Probability Problems

Mathematics is no more computation than typing is literature.
– John Allen Paulos

FSA Subject Test Mathematics Grade 7

Mean and Median

✏ Find Mean and Median of the Given Data.

1) 8, 7, 14, 4, 8

2) 14, 8, 25, 19, 16, 33, 11

3) 23, 18, 15, 12, 17

4) 34, 14, 10, 15, 6, 11

5) 10, 19, 6, 8, 32, 20, 17

6) 17, 26, 39, 69, 20, 6

7) 40, 38, 18, 11, 9, 2, 7, 32, 41

8) 24, 21, 31, 12, 33, 32, 22

9) 16, 14, 20, 41, 15, 20, 38, 4

10) 20, 20, 30, 18, 6, 28, 12, 46

11) 12, 7, 10, 11, 16, 22

12) 10, 29, 27, 12, 2, 15, 10, 3

✏ Calculate.

13) In a javelin throw competition, five athletics score 56, 34, 62, 23 and 19 meters. What are their Mean and Median? _____

14) Eva went to shop and bought 8 apples, 14 peaches, 6 bananas, 4 pineapples and 12 melons. What are the Mean and Median of her purchase? _____

15) Bob has 17 black pen, 19 red pen, 14 green pens, 20 blue pens and 5 boxes of yellow pens. If the Mean and Median are 19 respectively, what is the number of yellow pens in each box? _____

WWW.MathNotion.Com

FSA Subject Test Mathematics Grade 7

Mode and Range

✍ **Find Mode and Rage of the Given Data.**

1) 4, 3, 7, 3, 3, 4
   Mode: _____   Range: _____

2) 18, 18, 24, 26, 18, 8, 14, 22
   Mode: _____   Range: _____

3) 8, 8, 8, 16, 19, 22, 20, 9, 13
   Mode: _____   Range: _____

4) 24, 24, 14, 28, 20, 18, 20, 24
   Mode: _____   Range: _____

5) 6, 21, 27, 24, 27, 27
   Mode: _____   Range: _____

6) 21, 8, 8, 7, 8, 12, 10, 22, 18, 13
   Mode: _____   Range: _____

7) 7, 4, 4, 6, 13, 13, 13, 0, 2, 2
   Mode: _____   Range: _____

8) 5, 8, 5, 14, 12, 14, 3, 5, 18
   Mode: _____   Range: _____

9) 7, 7, 7, 12, 7, 3, 8, 16, 3, 17
   Mode: _____   Range: _____

10) 15, 15, 19, 16, 4, 16, 10, 15
    Mode: _____   Range: _____

11) 6, 6, 5, 6, 42, 13, 19, 2
    Mode: _____   Range: _____

12) 8, 8, 9, 8, 9, 4, 34, 22
    Mode: _____   Range: _____

✍ **Calculate.**

13) A stationery sold 12 pencils, 56 red pens, 24 blue pens, 20 notebooks, 12 erasers, 21 rulers and 11 color pencils. What are the Mode and Range for the stationery sells?

    Mode: _____   Range: _____

14) In an English test, eight students score 10, 15, 15, 18 18, 16, 15 and 15. What are their Mode and Range? _____

15) What is the range of the first 6 even numbers greater than 8?
    _____

WWW.MathNotion.Com

FSA Subject Test Mathematics Grade 7

Times Series

✎ Use the following Graph to complete the table.

Day	Distance (km)
1	
2	

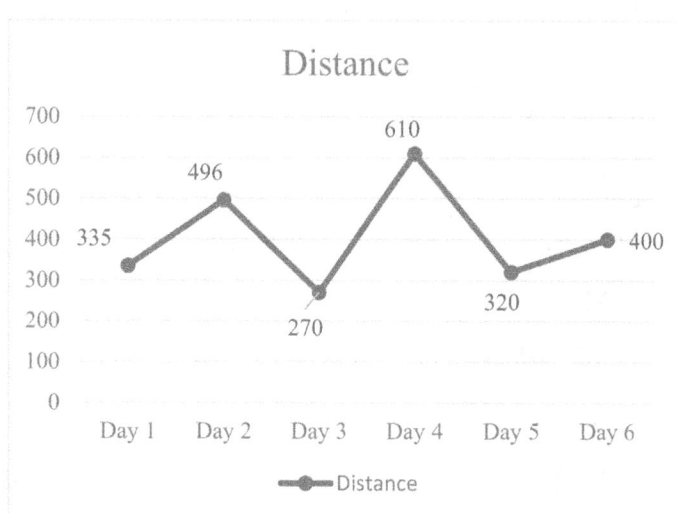

The following table shows the number of births in the US from 2007 to 2012 (in millions).

Year	Number of births (in millions)
2007	4.15
2008	3.70
2009	3.45
2010	3.20
2011	1.75
2012	2.98

Draw a Time Series for the table.

FSA Subject Test Mathematics Grade 7

Stem–and–Leaf Plot

✎ **Make stem ad leaf plots for the given data.**

1) 24, 26, 29, 20, 53, 27, 51, 55, 36, 21, 37, 30

   Stem | Leaf plot

2) 11, 59, 66, 14, 18, 19, 59, 65, 69, 61, 68, 65

   Stem | Leaf plot

3) 121, 55, 66, 54, 112, 128, 63, 125, 59, 123, 68, 119

   Stem | Leaf plot

4) 51, 32, 100, 56, 84, 36, 107, 56, 85, 39, 56, 106, 89

   Stem | Leaf plot

5) 33, 89, 19, 87, 81, 16, 11, 30, 86, 35, 17, 35, 13

   Stem | Leaf plot

6) 60, 92, 22, 25, 67, 93, 95, 62, 21, 64, 98, 29

   Stem | Leaf plot

WWW.MathNotion.Com

FSA Subject Test Mathematics Grade 7

Pie Graph

The circle graph below shows all Robert's expenses for last month. Robert spent $140 on his hobbies last month.

Answer following questions based on the Pie graph.

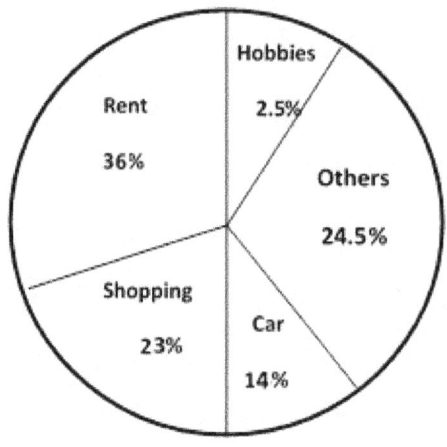

1) How much was Robert's total expenses last month? _____

2) How much did Robert spend on his car last month? _____

3) How much did Robert spend for shopping last month? _____

4) How much did Robert spend on his rent last month? _____

5) What fraction is Robert's expenses for his rent and car out of his total expenses last month? _____

WWW.MathNotion.Com

Probability Problems

✏ **Calculate.**

1) A number is chosen at random from 1 to 10. Find the probability of selecting number 6 or smaller numbers. _____

2) Bag A contains 18 red marbles and 6 green marbles. Bag B contains 16 black marbles and 8 orange marbles. What is the probability of selecting a green marble at random from bag A? What is the probability of selecting a black marble at random from Bag B? _____

3) A number is chosen at random from 1 to 20. What is the probability of selecting multiples of 4? _____

4) A card is chosen from a well-shuffled deck of 52 cards. What is the probability that the card will be a queen? _____

5) A number is chosen at random from 1 to 15. What is the probability of selecting a multiple of 3 or 5? _____

A spinner numbered 1–8, is spun once. What is the probability of spinning …?

6) an Odd number? _____   7) a multiple of 2? _____

8) a multiple of 5? _____   9) number 10? _____

FSA Subject Test Mathematics Grade 7

Answers of Worksheets

Mean and Median

1) Mean: 8.2, Median: 8
2) Mean: 18, Median: 16
3) Mean: 17, Median: 17
4) Mean: 15, Median: 12.5
5) Mean: 16, Median: 17
6) Mean: 29.5, Median: 23
7) Mean: 22, Median: 18
8) Mean: 25, Median: 24
9) Mean: 21, Median: 18
10) Mean: 22.5, Median: 20
11) Mean: 13, Median: 11.5
12) Mean: 13.5, Median: 11
13) Mean: 38.8, Median: 34
14) Mean: 8.8, Median: 8
15) 5

Mode and Range

1) Mode: 3, Range: 4
2) Mode: 18, Range: 18
3) Mode: 8, Range: 14
4) Mode: 24, Range: 14
5) Mode: 27, Range: 21
6) Mode: 8, Range: 15
7) Mode: 13, Range: 13
8) Mode: 5, Range: 15
9) Mode: 7, Range: 14
10) Mode: 15, Range: 15
11) Mode: 6, Range: 40
12) Mode: 8, Range: 30
13) Mode: 12, Range: 45
14) Mode: 15, Range: 8
15) 10

Time series

Day	Distance (km)
1	335
2	496
3	270
4	610
5	320
6	400

Stem–And–Leaf Plot

1)

Stem	leaf
2	0 1 4 6 7 9
3	0 6 7
5	1 3 5

2)

Stem	leaf
1	1 4 8 9
5	9 9
6	1 5 5 6 8 9

3)

Stem	leaf
5	4 5 9
6	3 6 8
11	2 9
12	1 3 5 8

FSA Subject Test Mathematics Grade 7

4)

Stem	leaf
3	2 6 9
5	1 6 6 6
8	4 5 9
10	0 6 7

5)

Stem	leaf
1	1 3 6 7 9
3	0 3 5 5
8	1 6 7 9

6)

Stem	leaf
2	2 1 5 9
6	0 2 4 7
9	2 3 5 8

Pie Graph

1) $5,600

2) $784

3) $1,288

4) $2,016

5) $\frac{1}{2}$

Probability Problems

1) $\frac{3}{5}$

2) $\frac{1}{4}, \frac{2}{3}$

3) $\frac{1}{4}$

4) $\frac{1}{13}$

5) $\frac{7}{15}$

6) $\frac{1}{2}$

7) $\frac{1}{2}$

8) $\frac{1}{8}$

9) 0

WWW.MathNotion.Com

FSA Subject Test Mathematics Grade 7

FSA Subject Test Mathematics Grade 7

Chapter 12 : FSA Math Practice Tests

Time to Test

Time to refine your skill with a practice examination.

Take two practice Grade 7 FSA Math Tests to simulate the test day experience.

After you've finished, score your test using the answer key.

Before You Start

- You'll need a pencil and scratch papers to take the test.

- For this practice test, don't time yourself. Spend time as much as you need.

- It's okay to guess. You won't lose any points if you're wrong.

- After you've finished the test, review the answer key to see where you went go.

Calculators are not permitted for students taking the

FSA Mathematics Grade 7.

Good Luck!

WWW.MathNotion.Com

FSA Subject Test Mathematics Grade 7

FSA GRADE 7 MAHEMATICS REFRENCE MATERIALS

Linear Equations

Slope-intercept form $\qquad y = mx + b$

Constant of proportionality $\qquad k = \frac{y}{x}$

Circumference

Circle $\qquad C = 2\pi r \qquad$ or $\qquad C = \pi d$

Area

Triangle $\qquad A = \frac{1}{2}bh$

Rectangle or Parallelogram $\qquad A = bh$

Trapezoid $\qquad A = \frac{1}{2}h(b_1 + b_2)$

Circle $\qquad A = \pi r^2$

Volume

Prism or cylinder $\qquad V = Bh$

Pyramid or Cone $\qquad V = \frac{1}{3}Bh$

Additional Information

Pi $\qquad \pi = 3.14 \qquad$ or $\qquad \pi = \frac{22}{7}$

Distance $\qquad d = rt$

Simple interest $\qquad I = prt$

Compound interest $\qquad I = p(1 + r)^t$

WWW.MathNotion.Com

FSA Subject Test Mathematics Grade 7

The Florida Standards Assessments
FSA Practice Test 1

Mathematics

GRADE 7

Released *Month Year*

FSA Subject Test Mathematics Grade 7

Session 1

- Calculators are NOT permitted for this practice test.
- Time for Session 1: 60 Minutes

FSA Subject Test Mathematics Grade 7

1) The price of a car was $50,000 in 2014, $41,000 in 2015 and $33,620 in 2016.

   What is the rate of depreciation of the price of car per year?

   A. 25 %

   B. 12 %

   C. 18 %

   D. 22 %

2) The square of a number is $\frac{25}{36}$. What is the cube of that number?

   A. $\frac{36}{25}$

   B. $\frac{216}{125}$

   C. $\frac{125}{216}$

   D. $\frac{15}{18}$

3) What is the value of x in the following equation?

   $$\frac{3}{7}x + \frac{1}{6} = \frac{5}{6}$$

   A. 14

   B. $\frac{14}{9}$

   C. $\frac{9}{14}$

   D. $\frac{1}{9}$

FSA Subject Test Mathematics Grade 7

4) Which of the following points lies on the line 6x + 7y = −4?

   A. (3, 5)

   B. (−4, 3)

   C. (−3, 2)

   D. (2, 3)

5) What is the surface area of the cylinder below?

   A. 64 π

   B. 80 π

   C. 32 π

   D. 40 π

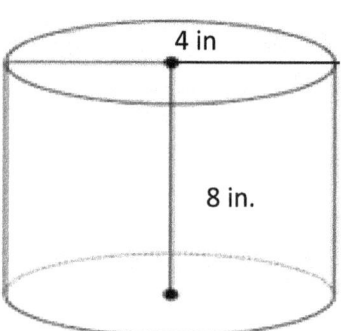

6) Anita's trick–or–treat bag contains 23 pieces of chocolate, 20 suckers, 21 pieces of gum, 26 pieces of licorice. If she randomly pulls a piece of candy from her bag, what is the probability of her pulling out a piece of sucker?

   A. $\frac{5}{9}$

   B. $\frac{1}{2}$

   C. $\frac{1}{9}$

   D. $\frac{2}{9}$

FSA Subject Test Mathematics Grade 7

7) Which of the following shows the numbers in?

$$\frac{8}{7}, 0.43, 69\%, \frac{2}{9}$$

  A. $69\%, 0.43, \frac{8}{7}, \frac{2}{9}$

  B. $69\%, 0.43, \frac{2}{9}, \frac{8}{7}$

  C. $0.43, 69\%, \frac{8}{7}, \frac{2}{9}$

  D. $\frac{8}{7}, 69\%, 0.43, \frac{2}{9}$

8) The mean of 28 test scores was calculated as 55. But it turned out that one of the scores was misread as 55 but it was 41. What is the correct mean of the data?

  A. 52.6

  B. 64

  C. 54.5

  D. 65.5

9) The width of a box is one third of its length. The height of the box is one third of its width. If the length of the box is 36cm, what is the volume of the box?

  A. 846 cm³

  B. 684 cm³

  C. 1,822 cm³

  D. 1,728 cm³

FSA Subject Test Mathematics Grade 7

10) In five successive hours, a car travels 51 km, 56 km, 58 km, 68 km and 60 km. In the next five hours, it travels with an average speed of 59 km per hour. Find the total distance the car traveled in 10 hours.

   A. 588 km

   B. 966 km

   C. 440 km

   D. 220 km

11) What is the slope of a line that is parallel to the line $12x - 6y = 18$?

   A. -3

   B. 2

   C. 6

   D. 8

12) In 1999, the average worker's income increased $9,000 per year starting from $53,000 annual salary. Which equation represents income greater than average? (I = income, x = number of years after 1999)

   A. I > 9,000 x + 53,000

   B. I > – 9,000 x + 53,000

   C. I < –9,000 x + 53,000

   D. I < 9,000 x – 53,000

13) Which of the following graphs represents the compound inequality $-9 \leq 7x - 9 < 5$?

A. 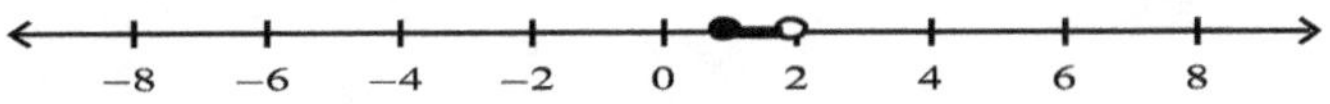

B. 

C. 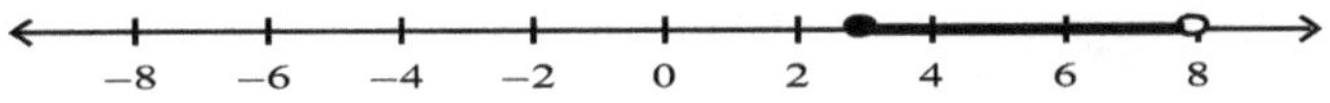

D. 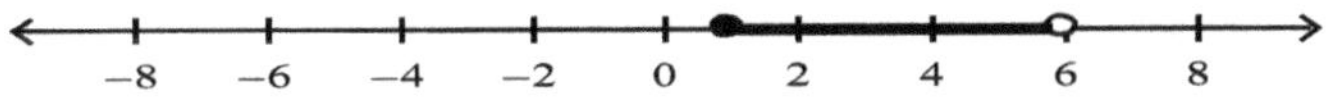

14) What is the value of the expression $3(2x + 3y) - (15 - 3x)^2$ when $x = 4$ and $y = -3$?

A. -12

B. 6

C. 24

D. 18

15) Two dice are thrown simultaneously, what is the probability of getting a sum of 8 or 10?

A. $\frac{5}{13}$

B. $\frac{1}{4}$

C. $\frac{2}{13}$

D. $\frac{2}{9}$

FSA Subject Test Mathematics Grade 7

Session 2

- Calculators are NOT permitted for this practice test.

- Time for Session 2: 60 Minutes

FSA Subject Test Mathematics Grade 7

16) Which graph corresponds to the following inequalities?

$$y \geq -3x + 6$$

$$-4x + 2y \leq 8$$

A.

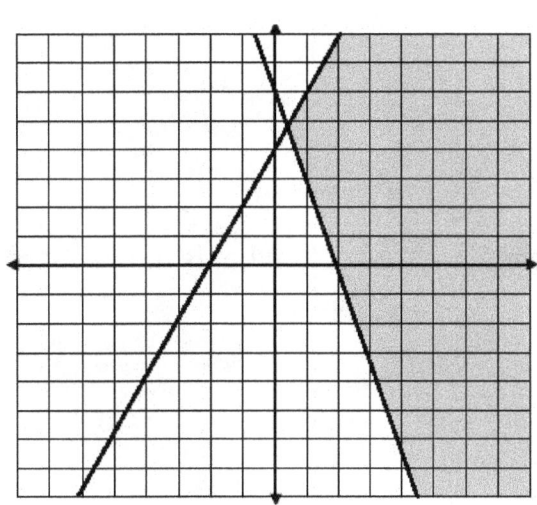

B.

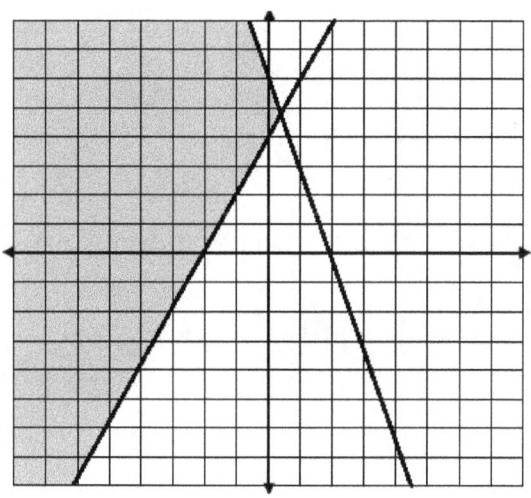

C.

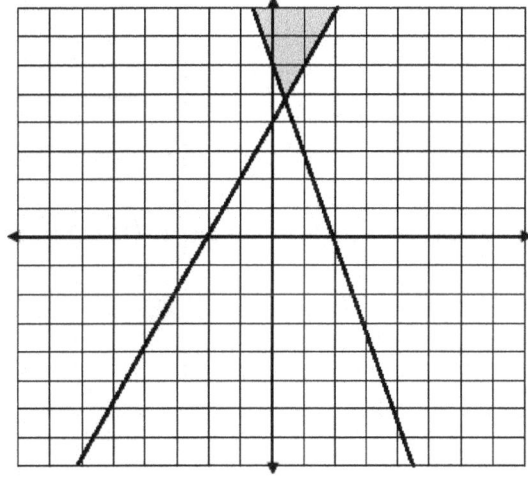

D.

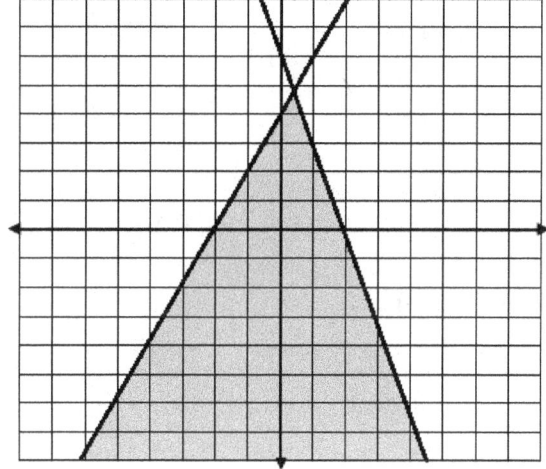

FSA Subject Test Mathematics Grade 7

17) A bank is offering 4.2% simple interest on a savings account. If you deposit $5,800 how much interest will you earn in five years?

   A. $812

   B. $1,218

   C. $1,820

   D. $2,436

18) A card is drawn at random from a standard 52–card deck, what is the probability that the card is of queen or aces?

   A. $\frac{2}{13}$

   B. $\frac{4}{13}$

   C. $\frac{5}{13}$

   D. $\frac{3}{52}$

19) How long does a 294–miles trip take moving at 28 miles per hour (mph)?

   A. 10 hours

   B. 10 hours and 30 minutes

   C. 10 hours and 20 minutes

   D. 10 hours and 50 minutes

FSA Subject Test Mathematics Grade 7

20) A shirt costing $750 is discounted 14%. After a month, the shirt is discounted another 14%. Which of the following expressions can be used to find the selling price of the shirt?

   A. (750) (0.14)

   B. (750) – 750 (0.14)

   C. (750) (0.14) – (750) (0.86)

   D. (750) (0. 86) (0.86)

21) Mr. Carlos family are choosing a menu for their reception. They have 8 choices of appetizers, 9 choices of entrees, 5 choices of cake. How many different menu combinations are possible for them to choose?

   A. 45

   B. 72

   C. 360

   D. 720

22) Six one – foot rulers can be split among how many users to leave each with $\frac{1}{7}$ of a ruler?

   A. 14

   B. 56

   C. 36

   D. 42

FSA Subject Test Mathematics Grade 7

23) What is the volume of a box with the following dimensions?

   Height = 4 cm   Width = 7 cm   Length = 11 cm

   A. $77 cm^3$

   B. $280\ cm^3$

   C. $308\ cm^3$

   D. $2,380\ cm^3$

24) What is the area of the shaded region?

   A. $54\ ft^2$

   B. $72\ ft^2$

   C. $144\ ft^2$

   D. $280\ ft^2$

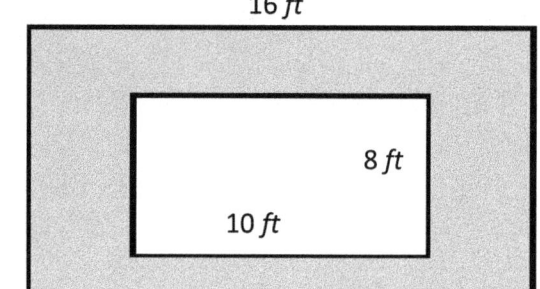

25) Mr. Jones saves $2,500 out of his monthly family income of $42,500. What fractional part of his income does he save?

   A. $\frac{1}{17}$

   B. $\frac{3}{17}$

   C. $\frac{1}{19}$

   D. $\frac{3}{19}$

FSA Subject Test Mathematics Grade 7

26) The ratio of boys and girls in a class is 8:7. If there are 90 students in the class, how many more boys should be enrolled to make the ratio 1:1?

   A. 6

   B. 12

   C. 42

   D. 48

27) When a number is subtracted from 110 and the difference is divided by that number, the result is 9. What is the value of the number?

   A. 44

   B. 33

   C. 22

   D. 11

28) The radius of the following cylinder is 7 inches, and its height is 13 inches. What is the surface area of the cylinder?

   A. 91π in^2

   B. 560π in^2

   C. 140π in^2

   D. 280π in^2

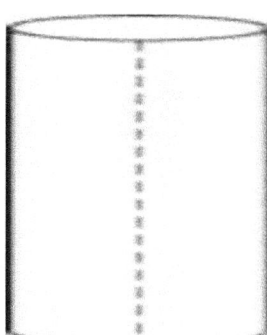

29) How many tiles of 14 cm^2 is needed to cover a floor of dimension 7 cm by 60 cm?

A. 40

B. 30

C. 20

D. 15

30) In two successive years, the population of a town is increased by 15% and 40%. What percent of its population is increased after two years?

A. 41

B. 61

C. 51

D. 11

FSA Subject Test Mathematics Grade 7

The Florida Standards Assessments

FSA Practice Test 2

Mathematics

GRADE 7

Released Month Year

FSA Subject Test Mathematics Grade 7

Session 1

- Calculators are NOT permitted for this practice test.
- Time for Session 1: 60 Minutes

FSA Subject Test Mathematics Grade 7

1) Right triangle ABC has two shorter legs of lengths 60 cm (AB) and 80 cm (AC).

   What is the length of the third side (BC)?

   A. 95cm

   B. 80 cm

   C. 100 cm

   D. 200cm

2) If $x = -7$, which equation is true?

   A. $x(3x + 12) = -50$

   B. $8(4 - x) = 88$

   C. $4(3x + 2) = -85$

   D. $11x - 4 = -76$

3) In a bag of small balls $\frac{1}{8}$ are black, $\frac{1}{16}$ are white, $\frac{1}{2}$ are red and the remaining 25 blue. How many balls are white?

   A. 8

   B. 10

   C. 5

   D. 3

4) A boat sails 15 miles south and then 36 miles east. How far is the boat from its start point?

   A. 48

   B. 40

   C. 39

   D. 42

5) Sophia purchased a sofa for $396. The sofa is regularly priced at $450. What was the percent discount Sophia received on the sofa?

   A. 2%

   B. 12%

   C. 92%

   D. 88%

6) The score of Elise was one fifth as that of Mia and the score of Stella was triple that of Mia. If the score of Stella was 90, what is the score of Elise?

   A. 12

   B. 18

   C. 6

   D. 36

FSA Subject Test Mathematics Grade 7

7) A bag contains 22 balls: eight green, four black, four blue, a brown, three red and two white. If 18 balls are removed from the bag at random, what is the probability that a brown ball has been removed?

   A. $\frac{9}{11}$

   B. $\frac{1}{6}$

   C. $\frac{2}{18}$

   D. $\frac{1}{22}$

8) A rope weighs 550 grams per meter of length. What is the weight in kilograms of 28.6 meters of this rope? (1 kilograms = 1000 grams)

   A. 0.157

   B. 1,573

   C. 15.73

   D. 157.3

9) What is the area of the shaded region?

   A. 105π in²

   B. 55π in²

   C. 51π in²

   D. 81π in²

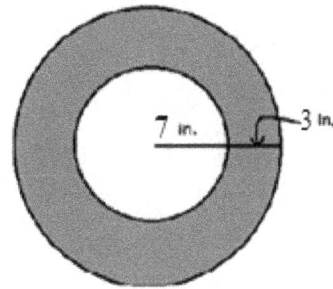

FSA Subject Test Mathematics Grade 7

10) A pizza cut into 12 parts. Elise and her sister Etta ordered two pizzas. Elise ate $\frac{5}{6}$ of her pizza and Etta ate $\frac{2}{3}$ of her pizza. What part of the two pizzas was left?

A. $\frac{1}{4}$

B. $\frac{3}{4}$

C. $\frac{7}{24}$

D. $\frac{5}{24}$

11) Simplify $4y^6(2x^6y)^4 =$

A. $32x^{10}y^{16}$

B. $32x^{20}y^{11}$

C. $64x^{24}y^{10}$

D. $64x^{10}y^{22}$

12) The marked price of a computer is D dollar. Its price decreased by 10% in January and later increased by 20% in February. What is the final price of the computer in D dollar?

A. 1.08 D

B. 0.92 D

C. 0.98 D

D. 1.02 D

FSA Subject Test Mathematics Grade 7

13) A $120 shirt now selling for $48 is discounted by what percent?

   A. 65%

   B. 60%

   C. 40%

   D. 35%

14) What is the median of these numbers? 17, 40, 28, 30, 37, 46, 23

   A. 37

   B. 30

   C. 23

   D. 46

15) From last year, the price of gasoline has increased from $1.50 per gallon to $2.10 per gallon. The new price is what percent of the original price?

   A. 60%

   B. 160%

   C. 140%

   D. 40%

16) The following trapezoid are similar. What is the value of x ?

   A. 7

   B. 14

   C. 12

   D. 16

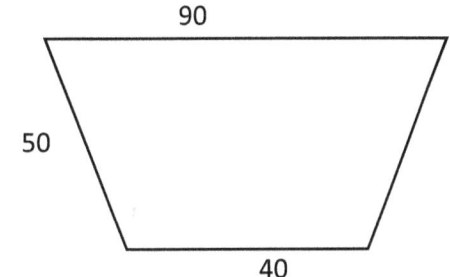
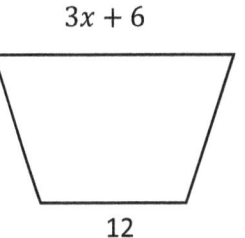

WWW.MathNotion.Com

FSA Subject Test Mathematics Grade 7

Session 2

- Calculators are NOT permitted for this practice test.
- Time for Session 2: 60 Minutes

FSA Subject Test Mathematics Grade 7

17) If 55% of a class are girls, and 34% of girls play tennis, what fraction of the class play tennis?

   A. 17.8%

   B. 8.17%

   C. 18.7%

   D. 28.7%

18) A chemical solution contains 14% alcohol. If there is 70 ml of alcohol, what is the volume of the solution?

   A. 150 ml

   B. 380 ml

   C. 500 ml

   D. 1,500 ml

19) The price of a laptop is decreased by 32% to $884. What is its original price?

   A. 860

   B. 1,300

   C. 2,600

   D. 1,600

FSA Subject Test Mathematics Grade 7

20) Last week 16,000 fans attended a football match. This week five times as many bought tickets, but one eighth of them cancelled their tickets. How many are attending this week?

   A. 80,000

   B. 70,000

   C. 10,000

   D. 90,000

21) 8 less than twice a positive integer is 56. What is the integer?

   A. 32

   B. 36

   C. 34

   D. 30

22) Four times the price of a laptop is equal to eight times the price of a computer. If the price of laptop is $800 more than the computer, what is the price of the computer?

   A. 600

   B. 1,200

   C. 400

   D. 800

23) Jason is 32 miles ahead of Joe running at 7 miles per hour and Joe is running at the speed of 15 miles per hour. How long does it take Joe to catch Jason?

   A. 8 hours

   B. 4 hours

   C. 9 hours

   D. 2 hours

24) 90 students took an exam and 18 of them failed. What percent of the students passed the exam?

   A. 20 %

   B. 10 %

   C. 80 %

   D. 90 %

25) What is the volume of the following square pyramid?

   A. 625 m³

   B. 125 m³

   C. 525 m³

   D. 155 m³

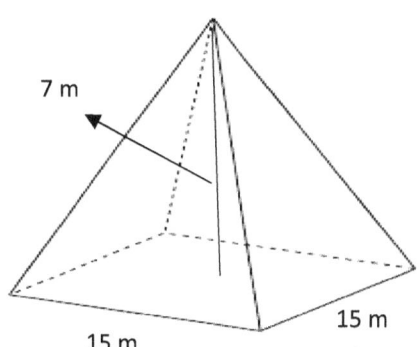

FSA Subject Test Mathematics Grade 7

26) Which of the following points lies on the line $3x + 7y = 5$?

   A. (2, 5)

   B. (−3, 2)

   C. (−1, 4)

   D. (4, 3)

27) An angle is equal to one fourth of its supplement. What is the measure of that angle?

   A. 36

   B. 30

   C. 45

   D. 20

28) 4.4 is what percent of 55?

   A. 5

   B. 8

   C. 10

   D. 2.5

29) What is the length of BC in the following figure if AB = 24, DF = 32 and BD = 84?

   A. 36

   B. 34

   C. 26

   D. 24

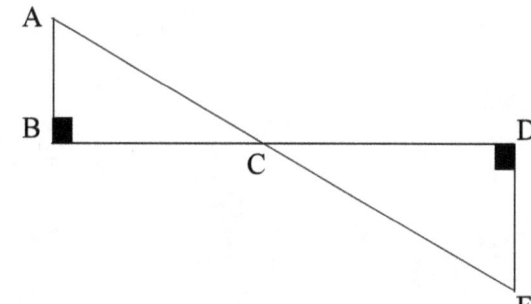

30) A bank is offering 2.2% simple interest on a savings account. If you deposit $36,000, how much interest will you earn in five years?

   A. $3,960

   B. $3,660

   C. $2,980

   D. $2,890

FSA Subject Test Mathematics Grade 7

FSA Subject Test Mathematics Grade 7

Chapter 13 : Answers and Explanations

FSA Practice Tests
Answer Key

❋ Now, it's time to review your results to see where you went wrong and what areas you need to improve!

SBAC - Mathematics

Practice Test - 1

1	C	11	B	21	C
2	C	12	A	22	D
3	B	13	B	23	C
4	C	14	A	24	C
5	D	15	D	25	A
6	D	16	A	26	A
7	D	17	B	27	D
8	C	18	A	28	D
9	D	19	B	29	B
10	A	20	D	30	B

Practice Test - 2

1	C	11	C	21	A
2	B	12	A	22	D
3	C	13	B	23	B
4	C	14	B	24	C
5	B	15	C	25	C
6	C	16	A	26	B
7	D	17	C	27	A
8	C	18	C	28	B
9	C	19	B	29	A
10	A	20	B	30	A

FSA Subject Test Mathematics Grade 7

FSA Subject Test Mathematics Grade 7

Practice Test 1
SBAC - Mathematics
Answers and Explanations

1) Answer: C

Use this formula: Percent of Change: $\frac{\text{New Value} - \text{Old Value}}{\text{Old Value}} \times 100\%$

$\frac{41,000 - 50,000}{50,000} \times 100\% = -18\%$ and $\frac{33,620 - 41,000}{41,000} \times 100\% = -18\%$

2) Answer: C

The square of a number is $\frac{25}{36}$, then the number is the square root of $\frac{25}{36}$: $\sqrt{\frac{25}{36}} = \frac{5}{6}$;

The cube of the number is: $(\frac{5}{6})^3 = \frac{125}{216}$

3) Answer: B

$\frac{3}{7}x + \frac{1}{6} = \frac{5}{6} \Rightarrow \frac{3}{7}x = \frac{4}{6} = \frac{2}{3} \Rightarrow x = \frac{7}{3} \times \frac{2}{3} \Rightarrow x = \frac{14}{9} = 1\frac{5}{9}$

4) Answer: C

Input $(-3, 2)$ in the $6x + 7y = -4$ formula instead of x and y. So, we have: $6(-3) + 7(2) = -4 \Rightarrow -18 + 14 = -4$

5) Answer: D

Surface Area of a cylinder = $2\pi r(r + h)$,

The radius of the cylinder is 2 (4 ÷ 2) inches, and its height is 8 inches. Therefore,

Surface Area of a cylinder = $2\pi(2)(2 + 8) = 40\pi$

6) Answer: D

Probability = $\frac{\text{number of desired outcomes}}{\text{number of total outcomes}} = \frac{20}{23 + 20 + 21 + 26} = \frac{20}{90} = \frac{2}{9}$

7) Answer: D

Change the numbers to decimal and then compare.

$\frac{2}{9} = 0.22$;   0.43;   69% = 0.69; $\frac{8}{7} = 1.14$

$\Rightarrow$ Therefore $\frac{8}{7} > 69\% > 0.43 > \frac{2}{9}$

FSA Subject Test Mathematics Grade 7

8) Answer: C

average (mean) = $\dfrac{\text{sum of terms}}{\text{number of terms}} \Rightarrow 55 = \dfrac{\text{sum of terms}}{28} \Rightarrow$ sum = $55 \times 28 = 1{,}540$

The difference of 55 and 41 is 14. Therefore, 14 should be subtracted from the sum.

$1{,}540 - 14 = 1{,}526$

mean = $\dfrac{\text{sum of terms}}{\text{number of terms}} \Rightarrow$ mean = $\dfrac{1{,}526}{28} = 54.5$

9) Answer: D

If the length of the box is 36, then the width of the box is one-third of it, 12, and the height of the box is 4 (one-third of the width). The volume of the box is:

$V = l \times w \times h = (36) \times (12) \times (4) = 1{,}728$

10) Answer: A

Add the first 5 numbers. $51 + 56 + 58 + 68 + 60 = 293$

To find the distance traveled in the next 5 hours, multiply the average by number of hours.

Distance = Average × Rate = $59 \times 5 = 295$, Add both numbers: $293 + 295 = 588$

11) Answer: B

The equation of a line in slope intercept form is: $y = mx + b$

Solve for y.

$12x - 6y = 18 \Rightarrow -6y = 18 - 12x \Rightarrow y = (18 - 12x) \div (-6) \Rightarrow y = 2x - 3$

The slope of this line is 2. Parallel lines have same slopes.

12) Answer: A

Let x be the number of years. Therefore, \$9,000 per year equals $9{,}000x$.

starting from \$53,000 annual salary means you should add that amount to $9{,}000x$.

Income more than that is: I > $9{,}000x + 53{,}000$.

13) Answer: B

Solve for x.

$-9 \leq 7x - 9 < 5 \Rightarrow$ (add 9 all sides)

$-9 + 9 \leq 7x - 9 + 9 < 5 + 9$

$\Rightarrow 0 \leq 7x < 14 \Rightarrow$ (divide all sides by 7) $0 \leq x < 2$

FSA Subject Test Mathematics Grade 7

x is between 0 and 2. Choice B represent this inequality.

14) Answer: A

When $x = 4$ and $y = -3$,

Substitute: $3(2x + 3y) - (15 - 3x)^2 = 3(2(4) + 3(-3)) - (15 - 3(4))^2$

$= 3(8 - 9) - (15 - 12)^2 = -3 - 9 = -12$

15) Answer: D

The options to get sum of 8: (2 & 6) and (6 & 2), (3 & 5) and (5 & 3) and (4 & 4), so we have 5 options.

The options to get sum of 10 (4 & 6), (6 & 4) and (5 & 5), we have 3 options.

To get the sum of 8 or 10 for two dice, we have 8 options: $5 + 3 = 8$

Since, we have $6 \times 6 = 36$ total options, the probability of getting a sum of 8 and 10 is 8 out of 36 or $\frac{8}{36} = \frac{2}{9}$.

16) Answer: A

For each option, choose a point in the solution part and check it on both inequalities.

$$y \geq -3x + 6$$

$$-4x + 2y \leq 8$$

A. Point (3, 1) is in the solution section. Let's check the point in both inequalities.

$1 \geq -9 + 6$, It works. $-4(3) + 2(1) \leq 8 \Rightarrow -10 \leq 8$, it works (this point works in both)

B. Let's choose this point $(-3, 2)$; $2 \geq -3(-3) + 6$, That's not true!

C. Let's choose this point $(1, 7)$; $7 \geq -3(1) + 6$, It works.

$-4(1) + 2(7) \leq 8$, That's not true!

D. Let's choose this point $(0, 0)$; $0 \geq -3(0) + 6$, That's not true!

17) Answer: B

Use simple interest formula:

$I = prt$ (I = interest, p = principal, r = rate, t = time)

$I = (5,800)(0.042)(5) = 1,218$

FSA Subject Test Mathematics Grade 7

18) Answer: A

The probability of choosing a queen or aces is $\frac{1}{13} + \frac{1}{13} = \frac{2}{13}$

19) Answer: B

Use distance formula:

Distance = Rate × time ⇒ 294 = 28 × T, divide both sides by 28.

$\frac{294}{28} = T \Rightarrow T = 10.5$ hours.

Change hours to minutes for the decimal part.

0.5 hours = 0.5 × 60 = 30 minutes.

20) Answer: D

To find the discount, multiply the number by (100% – rate of discount).

Therefore, for the first discount we get: (750)(100% – 14%) = (750)(0.86)

For the next 14% discount: (750)(0.86)(0.86)

21) Answer: C

To find the number of possible outfit combinations, multiply number of options for each factor: 8 × 9 × 5 = 360

22) Answer: D

$6 \div \frac{1}{7} = 42$

23) Answer: C

Volume of a box = length × width × height = 11 × 7 × 4 = 308

24) Answer: C

Use the area of rectangle formula ($s = a \times b$).

To find area of the shaded region subtract the smaller rectangle from bigger rectangle.

$S_1 - S_2 = (16ft \times 14t) - (8ft \times 10ft)$

$\Rightarrow S_1 - S_2 = 224ft^2 - 80ft^2 = 144ft^2$

25) Answer: A

2,500 out of 42,500 equals to $\frac{2,500}{42,500} = \frac{25}{425} = \frac{1}{17}$

FSA Subject Test Mathematics Grade 7

26) Answer: A

The ratio of boy to girls is 8:7. Therefore, there are 8 boys out of 15 students.

To find the answer, first divide the total number of students by 15, then multiply the result by 8.

$90 \div 15 = 6 \Rightarrow 6 \times 8 = 48$

There are 48 boys and 42 (90 – 48) girls. So, 6 more boys should be enrolled to make the ratio 1:1

27) Answer: D

Let the number be x. Then:

$\frac{110-x}{x} = 9 \to 9x = 110 - x \to 10x = 110 \to x = 11$

28) Answer: D

Surface Area of a cylinder = $2\pi r (r + h)$,

The radius of the cylinder is 7 inches, and its height is 13 inches. π is about 3.14. Then:

Surface Area of a cylinder = $2 (\pi) (7) (7 + 13) = 280 \pi$

29) Answer: B

The area of the floor is: 7 cm × 60 cm = 420 cm²

The number of tiles needed = 420 ÷ 14 = 30

30) Answer: B

The population is increased by 15% and 40%. 15% increase changes the population to 115% of original population.

For the second increase, multiply the result by 140%.

(1.15) × (1.40) = 1.61 = 161%

61 percent of the population is increased after two years.

FSA Subject Test Mathematics Grade 7

FSA Subject Test Mathematics Grade 7

Practice Test 2
SBAC - Mathematics
Answers and Explanations

1) Answer: C

Use Pythagorean Theorem: $a^2 + b^2 = c^2$

$60^2 + 80^2 = c^2 \Rightarrow 3{,}600 + 6{,}400 = c^2 \Rightarrow c^2 = 10{,}000 \Rightarrow c = 100$

2) Answer: B

Only option B is correct. Other options don't work in the equation.

$8(4 - (-7)) = 88$

3) Answer: C

Let x be the number of balls. Then:

$\frac{1}{8}x + \frac{1}{16}x + \frac{1}{2}x + 25 = x$

$(\frac{1}{8} + \frac{1}{16} + \frac{1}{2})x + 25 = x$

$(\frac{11}{16})x + 25 = x \Rightarrow x = 80$

In the bag of small balls $\frac{1}{16}$ are white, then: $\frac{80}{16} = 5$

There are 8 white balls in the bag.

4) Answer: C

Use the information provided in the question to draw the shape.

Use Pythagorean Theorem: $a^2 + b^2 = c^2$

$15^2 + 36^2 = c^2 \Rightarrow 225 + 1{,}296 = c^2 \Rightarrow 1{,}521 = c^2$

$\Rightarrow c = 39$

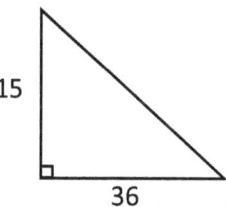

5) Answer: B

The question is this: 396 is what percent of 450?

Use percent formula: part = $\frac{\text{percent}}{100}$ × whole

$396 = \frac{\text{percent}}{100} \times 450 \Rightarrow 396 = \frac{\text{percent} \times 450}{100} \Rightarrow 39{,}600 = \text{percent} \times 450$

$\Rightarrow \text{percent} = \frac{39{,}600}{450} = 88$

FSA Subject Test Mathematics Grade 7

396 is 88 % of 450.

Therefore, the discount is: 100% – 88% = 12%

6) Answer: C

If the score of Stella was 90, then the score of Mia is 30. Since, the score of Elise was half as that of Mia, therefore, the score of Elise is 6.

7) Answer: D

If 18 balls are removed from the bag at random, there will be one ball in the bag. The probability of choosing a brown ball is 1 out of 22. Therefore, the probability of not choosing one brown ball is 18 out of 22 and the probability of having not brown ball after removing 18 balls is the same.

8) Answer: C

The weight of 28.6 meters of this rope is: 28.6 × 550 g = 15,730g

1kg = 1,000 g, therefore, 15,730g g ÷ 1,000 = 15.73 kg

9) Answer: C

To find the area of the shaded region, find the difference of the area of two circles. (S_1: the area of bigger circle. S_2: the area of the smaller circle)

Use the area of circle formula. $S = \pi r^2$

$$S1 - S2 = \pi(7+3)^2 - \pi(7in)^2 \Rightarrow S1 - S2 = \pi 100 in^2 - \pi 49\ in^2$$

$$\Rightarrow S1 - S2 = 51\pi\ in^2$$

10) Answer: A

Elise ate $\frac{5}{6}$ of 12 parts of his pizza that it means 10 parts out of 12 parts ($\frac{5}{6}$ of 12 parts = $x \Rightarrow x = 10$) and left 2 parts.

Etta $\frac{2}{3}$ of 12 parts of her pizza that it means 8 part out of 12 parts ($\frac{2}{3}$ of 12 parts = $x \Rightarrow x = 8$) and left 4 parts.

Therefore, they ate (10 + 8) parts out of (12 + 12) parts of their pizza and left (2 + 4) parts out of (12 + 12) parts of their pizza. It means: $\frac{6}{24}$

After simplification we have: $\frac{1}{4}$

FSA Subject Test Mathematics Grade 7

11) Answer: C

Simplify: $4y^6(2x^6y)^4 = 4y^6(16x^{24}y^4) = 64x^{24}y^{10}$

12) Answer: A

To find the discount, multiply the number by (100% – rate of discount).

Therefore, for the first discount we get: (D) (100% – 10%) = (D) (0.90) = 0.90 D

For increase of 20%: (0.9 D) (100% + 20%) = (0.90 D) (1.20) = 1.08 D = 108% of D

13) Answer: B

Use the formula for Percent of Change.

$\frac{\text{New Value} - \text{Old Value}}{\text{Old Value}} \times 100\%$

$\frac{48-120}{120} \times 100\% = -60\%$

(negative sign here means that the new price is less than old price).

14) Answer: B

Write the numbers in order: 17, 23, 28, 30, 37, 40, 46.

Since we have 7 numbers (7 is odd), then the median is the number in the middle, which is 30.

15) Answer: C

The question is this: 2.10 is what percent of 1.50?

Use percent formula: $\text{part} = \frac{\text{percent}}{100} \times \text{whole} \Rightarrow \frac{\text{percent}}{100} \times 1.50$

$\Rightarrow 2.10 = \frac{\text{percent} \times 1.50}{100} \Rightarrow 210 = \text{percent} \times 1.50 \Rightarrow \text{percent} = \frac{210}{1.50} = 140$

16) Answer: A

Write the ratio and solve for x.

$\frac{90}{40} = \frac{3x+6}{12} \Rightarrow 40(3x+6) = 90 \times 12 \Rightarrow 3x+6 = 1{,}080 \div 40 \Rightarrow 3x+6 = 27 \Rightarrow x = 7$

17) Answer: C

Let x be the number of students in the class.

55 % of x = girls

34 % of girls = tennis player; Find 34% of 55%. Then:

$34\% \text{ of } 55\% = 0.34 \times 0.55 = 0.187 = 18.7\%$

FSA Subject Test Mathematics Grade 7

18) Answer: C

14% of the volume of the solution is alcohol. Let x be the volume of the solution.

Then: 14% of x = 70 ml $\Rightarrow$ 0.14 x = 70 $\Rightarrow$ x = 70 ÷ 0.14 = 500

19) Answer: B

Let x be the original price.

If the price of a laptop is decreased by 32% to $884, then:

68 % $of\ x = 884 \Rightarrow 0.68x = 884 \Rightarrow x = 884 ÷ 0.68 = 1,300$

20) Answer: B

Five times of 16,000 is 80,000. One eighth of them cancelled their tickets.

One eighth of 80,000 equal 10,000 ($\frac{1}{8}$ × 80,000 = 10,000).

70,000 (80,000 – 10,000 = 70,000) fans are attending this week

21) Answer: A

Let x be the integer. Then: $2x - 8 = 56$

Add 8 both sides: $2x = 64$

Divide both sides by 2: $x = 32$

22) Answer: D

Let L be the price of laptop and C be the price of computer.

$4(L) = 8(C)$ and $L = \$800 + C$

Therefore, $4(\$800 + C) = 8C \Rightarrow \$3,200 + 4C = 8C \Rightarrow C = \$800$

23) Answer: B

The distance between Jason and Joe is 32 miles. Jason running at 7 miles per hour and Joe is running at the speed of 15 miles per hour. Therefore, every hour the distance is 8 miles less. $32 ÷ 8 = 4$

24) Answer: C

The failing rate is 18 out of 90 = $\frac{18}{90}$

Change the fraction to percent: $\frac{18}{90} \times 100\% = 20\%$

20 percent of students failed. Therefore, 80 percent of students passed the exam.

FSA Subject Test Mathematics Grade 7

25) Answer: C

Use the volume of square pyramid formula.

$V = \frac{1}{3}a^2 h \Rightarrow V = \frac{1}{3}(15m)^2 \times 7m \Rightarrow V = 525 m^3$

26) Answer: B

Input the points instead of x and y in the formula. Only option B works in the equation.

$3x + 7y = 5 \Rightarrow 3(-3) + 7(2) = 5$

27) Answer: A

The sum of supplement angles is 180. Let x be that angle. Therefore,

$x + 4x = 180$

$5x = 180$, divide both sides by 5: $x = 36$

28) Answer: B

x % of 55 is 4.4, then: x % 55 = 4.4 $\Rightarrow 0.55x = 4.4$

$\Rightarrow x = 4.4 \div 0.55 = 8$

29) Answer: A

Two triangles $\triangle ABC$ and $\triangle CD$ are similar. Then:

$\frac{AB}{DF} = \frac{BC}{CD} \rightarrow \frac{24}{32} = \frac{x}{84-x} \rightarrow 2{,}016 - 24x = 32x \rightarrow 56x = 2{,}016 \rightarrow x = 36$

30) Answer: A

Use simple interest formula: $I = prt$ (I = interest, p = principal, r = rate, t = time)

$I = (36{,}000)(0.022)(5) = 3{,}960$

"End"

www.ingramcontent.com/pod-product-compliance
Lightning Source LLC
Chambersburg PA
CBHW080440110426
42743CB00016B/3223